AF766340

Who Owns the Moon?

SPACE REGULATIONS LIBRARY

VOLUME 4

EDITORIAL BOARD

Managing Editor

PROF. R. JAKHU, *Institute of Air and Space Law, McGill University, Montreal, Canada*

MEMBERS

M. DAVIS, *Ward & Partners, Adelaide, Australia*
S. LE GOUEFF, *Le Goueff Law Office, Luxembourg*
P. NESGOS, *Milbank, Tweed, Hadley & McCloy, New York, U.S.A.*
S. MOSTESHAR, *Chambers of Sa'id Mosteshar, London, U.K. & Mosteshar Mackenzie, California, U.S.A.*
L. I. TENNEN, *Law Offices of Sterns and Tennen, Phoenix, Arizona, U.S.A.*

For other titles published in this series, go to
www.springer.com/series/6573

Virgiliu Pop

Who Owns the Moon?

Extraterrestrial Aspects of Land and Mineral Resources Ownership

Springer

Virgiliu Pop
Romanian Space Agency
Str. Mendeleev 21-25
Bucuresti 010362
Romania
virgiliu.pop@gmail.com

ISBN: 978-1-4020-9134-6 e-ISBN: 978-1-4020-9135-3

DOI 10.1007/978-1-4020-9135-3

Library of Congress Control Number: 2008935661

© Springer Science+Business Media B.V. 2009
No part of this work may be reproduced, stored in a retrieval system, or transmitted
in any form or by any means, electronic, mechanical, photocopying, microfilming, recording
or otherwise, without written permission from the Publisher, with the exception
of any material supplied specifically for the purpose of being entered
and executed on a computer system, for exclusive use by the purchaser of the work.

Printed on acid-free paper

9 8 7 6 5 4 3 2 1

springer.com

Introduction

This work investigates the permissibility and viability of property rights on the celestial bodies, particularly the extraterrestrial aspects of land and mineral resources ownership. In lay terms, it aims to find an answer to the question "Who owns the Moon?"

The first chapter critically analyses and dismantles with legal arguments the issue of sale of extraterrestrial real estate, after having perused some of the trivial claims of celestial bodies ownership. The only consequence these claims have on the plane of space law is to highlight the need for a better regulation of extraterrestrial landed property rights.

Next, the book addresses the apparent silence of the law in the field of extraterrestrial landed property, scrutinizing whether the factual situation on the extraterrestrial realms calls for legal regulations. The sources of law are examined in their dual dimension – that is, the facts that have caused and shaped the law of extraterrestrial real estate, and the norms which express this law. It is found that the norms and rules regarding property rights in the celestial realms are rather limited, failing to define basic concepts such as celestial body.

The following chapter examines precisely this issue, pondering whether asteroids and comets are immovable land-like territorial extensions that cannot be legally appropriated, or floating movable goods, capable of being captured and reduced into private ownership. The employment of the spatialist and functionalist approaches, the use of the criterion of actual movability from orbit by human action, and original theories such as the analogy between the legal status of asteroids and icebergs, are considered, concluding that some extraterrestrial resources are not, legally speaking, celestial bodies.

An examination follows of the relationship between appropriation under international law, and civil law appropriation, namely whether the non-appropriation principle in the Outer Space Treaty on the international plane, results also in the prohibition of the appropriation of the celestial bodies on the private property rights plane. It is offered that, while appropriation of land can exist outside the sphere of sovereignty, its survival is dependent upon backing from a sovereign entity, yet such endorsement would be unlawful as a means of national appropriation.

The next chapter answers to the main question of this book, offering that *de lege lata* the extraterrestrial realms, as a commons, belong to "everybody and nobody".

The contents of *Res communis* is duly scrutinized. *Res publica* is then presented together with the "Public Trust Doctrine". The commons regime currently accepted by most space actors is being challenged on two fronts. On the left, the adepts of the Common Heritage of Mankind paradigm have enacted the Moon Agreement, effectively planting the Marxist standard in the lunar soil. The egalitarian regime of extraterrestrial resource development would sanction the culture of entitlement, favouring a "fair outcome" over "fair process". Such an approach has failed on earth and is poised to fail in outer space, as argued in the following chapter.

On the right, the adepts of the "frontier paradigm" seek to promote individualism, competition, economic liberty, efficiency and laissez-faire economics – all linked to a privatization of the international public domain. The frontier paradigm has proven its worth on our planet, and it most likely will do so in the extraterrestrial realms. Property rights are a useful engine and, in al likelihood, a precondition for pushing forward the development of the extraterrestrial realms. Securing property rights would be more beneficial to humankind, compared to the alternative of keeping the extraterrestrial realms undeveloped.

The last chapter addresses the lex lata status of materials extracted from the Moon, these pertaining to the legal category of movable goods. Several viewpoints are presented, the conclusion being the permissibility, under the current regulations, of extracting and appropriating extraterrestrial material – be it as a scientific sample or as a commercial commodity.

Contents

Justification

In 1964, when presenting the students of The Hague Academy of International Law with a space law course, Judge Manfred Lachs (1964, p. 7) opened his lecture by rhetorically asking:

> Need I apologise for my choice of subject? Some may say it belongs to the realm of exotics of law. Some may ask: Why deal with issues so remote when there are so many much closer to us still awaiting a solution? Why reach so far?

These words were maybe echoing the comments of a newspaper published in 1956, in which the jurists present at the Astronautical Congress of Rome were criticised for "discussing the law to be applicable to the Martian canals, while for the time being the legal status of the Suez Canal was causing grave difficulties" (Gal, 1969, p. 201). In 1957, the Archbishop of Canterbury commented that "[t]he only people who are interested in this space business are people who have nothing better to think of, poor fellows". Not long after, an article published in The Economist (quoted in Goedhuis, 1978, p. 576) described the space lawyers as the "lunatic fringe of the profession".

Since then, humankind has sent its robotic messengers to the endless realms of the extraterrestrial space, and has walked on the Moon. Small steps have become giant leaps out of the cradle. Views have changed too: in 1978, Stephen T. Cheston (quoted in Moore, 1978, p. 269) was remarking that "[t]he future of space and the associated body of law are subjects of almost entrancing qualities; they spark the imagination, challenge the intellect and demand a level of wisdom normally reserved to those who live on Mount Olympus". One year later, Adrian Bueckling (1979, p. 18) warned that "[a]ttempts to develop and establish legal rules governing a vast and extremely complex subject matter like space exploration are rather like trying to hack down Mount Everest with a blunt kitchen knife".

We neither apologise for our choice of subject, nor would we ask for the compassion reserved for the ones not having anything better to think of. We only hope that it will be taken into consideration that this work was not written by an Olympian, but rather by an Everest-hacking Promethean.

Chapter 1
Is the Moon for Sale?[1]

You all think you own the moon, just because you never see anybody else looking at it

Orson Scott Card (1995)

What good is the Moon if you can't buy or sell it?

Ivan F. Boesky

1.1 Introduction

Once upon a time, a young boy left Asteroid B-612 and embarked on a journey that would take him across the solar system, fuelled by the imagination of Antoine de Saint-Exupéry. In his drift, the Little Prince sets on worlds inhabited by characters as diverse as a vain man, a geographer, and a businessman purporting to own hundreds of millions of stars. Asked by the Little Prince how he can own 501,622,731 stars, the businessman explains:-

> [T]hey are mine, because I was the first person to think of it ... When you find an island that is nobody's, it's yours. When you are the first one to have an idea, you copyright it: it's yours. And I own the stars, because nobody before me thought to own them (Exupéry 1946).

Saint-Exupéry's tale carries the reader not only across space, but across time as well, into a world where the orb-owning entrepreneur came to life.

For many years, specialists in space law bled ink over the question "who owns the Moon". Had they perused the media on a slow news day, the scholars could have learnt that the celestial satellite has been claimed, over and over again, by "inspired" individuals. Throughout the ages, many a dreamer would glimpse at the Moon and, struck by the thought that this is unclaimed property, would call the alien orb one's own.

During the last decade particularly, the fourth estate has been very enthusiastic in reporting on the extraordinary story of a man in the United States who "owns" the Moon and much of the Solar System, and who is more than happy to share his

---

[1]An extended form of this chapter has been published as a monograph – Pop, V., 2006. *Unreal estate: the Men Who Sold the Moon*, Exposure Publishing. Ideas from this chapter were published as Pop, V., 2001. The men who sold the Moon: science fiction or legal nonsense? *Space Policy*, Vol. 17, p. 195; Pop, V., 2003. Dennis Hope, the Masai, and the Moon. *Space Daily*, November 20; Pop, V., 2004. Extraterrestrial real estate: debunking the myth. *The 47th Colloquium on the Law of Outer Space*; Pop, V., 2006. The Nation of Celestial Space. *Space Policy*, Vol. 22, p. 205.

V. Pop, *Who Owns the Moon?,* Space Regulations Library Series 4,
DOI 10.1007/978-1-4020-9135-3_1, © Springer Science+Business Media B.V. 2009

extraterrestrial wealth with everybody – for a fee. By December 2005, the "Lunar Embassy", owned by Dennis M. Hope, boasted 3.6 million extraterrestrial "property owners" in 181 countries. Throughout the 28 years of its existence, the company has catered for people from all walks of life, such as attorneys, doctors, educators, but also for hundreds of celebrities counting Hollywood stars, politicians from many countries, and space travellers (Dennis Hope, personal communication, December 12, 2005). The advent of the global communications era had ushered an otherwise obscure novelty business into an undeserved spotlight, making it the leader of the extraterrestrial real estate business. Yet, unbeknown to many of its clients, the practice of selling "unreal estate" has been around for a long time.

The claim that the Moon belongs to a man self-styled "The Head Cheese" may be trivial; one cannot however ignore the seriousness of the problem behind this humorous mask. The regulation of property in Outer Space and on the Celestial Bodies is rather overlooked by the law and, where norms exist, they are far from being precise. In an era where a return to the Moon is being envisaged by NASA and private enterprise is poised to play a major role in opening the high frontier, the question of extraterrestrial ownership is not a hollow exercise.

While in the following chapters the importance and problematic nature of property rights in outer space will be outlined, one needs first to critically analyse and dismantle the issue of so-called "extraterrestrial real estate", after having outlined some of the trivial claims of celestial bodies ownership.

1.2 The Trivial Issue: "Extraterrestrial Real Estate"

For eons, the Moon has been the symbol of supreme desire. Asking for the Moon meant asking for the impossible. Owning the Moon and the stars was both the ultimate want and the ultimate folly. And, proven that a fool and his money are easily parted, there ought to arise people eager to exploit this weakness.

In the United States, lunar real estate was being peddled as early as the 1890s, when, according to journalist R. D. Whytock (quoted in Reno Evening Gazette, 1929), "unscrupulous money-gleaners" from New York were known for selling "lots on Luna ... and scores of other commodities that would appeal to the credulous public". And in 1910, lecturer Frank Dixon (quoted in The Daily Northwestern, 1910) was complaining that "there is no legal relief against the bunco steerer. Any man can organize a real estate company to sell lots on the planet Mars and get a charter in the state of New Jersey."

On June 15th, 1936, A. Dean Lindsay of Ocilla, Georgia, presented himself before a Pittsburgh Notary Public and made original claims for:-

> [a]ll of the property known as planets, islands-of-space or other matter, henceforth to be known as "A.D. Lindsay's archapellago" [*sic*] located in all the region visible (by any means) ... from the city of Ocilla, Ga, together with all ... matter (except this world ...) visible from any other planet, island-of-space or other matter.

In a letter sent to the Clerk of Superior Court in Ocilla, Lindsay said he realized these great holdings have no private owner, hence he decided to lay claim to them so that henceforth they will always have one. Accompanying the letter were the deeds sent for record, together with the due amount for recording them (Ocilla Star, 1937). And recorded they were on June 28th, in Deed Book 11, pp. 28 and 29, at Irwin County Courthouse in Ocilla.

Within a year, Lindsay caught the attention of the public. Over the radio, he was introduced as "the richest man in all the history of time". People desiring to buy the Moon, a star, or a constellation sent letters to him from various parts of the world, yet were met with refusal (Coffman, 1938, p. 12). The reason for keeping hold of his extraterrestrial goods was the commercial value Lindsay saw in them, and his plans for their exploitation (Ocilla Star, 1937). Indeed, Lindsay was careful to claim, for each set of celestial bodies registered,

> all ... natural ... improvements, ways, waters, water courses, rights, liberties, privileges, hereditaments and appurtenances ... and the revisions and remainders, rents, issues and profits thereof; and all the estate, right, title, interest, property, claim and demand whatsoever, in law, equity or otherwise, howsoever, of, in and to the same and every part thereof.

While Lindsay claimed the "islands-of-space", he did not care to grasp the space surrounding them. This would be accomplished over a decade later by industrialist James Thomas Mangan. Narrates Edwin A. Lotko (1968):-

> On that memorable day of the universe, December 20, 1948, at the stroke of midnight, after indefatigable research, James Thomas Mangan, standing high atop of the City of Chicago, reached out and seized all space in the sky in all directions away from the Earth as the complete possession and domain of the new sovereign Nation of Celestial Space.

Several factors contributed to the establishment of the Nation of Celestial Space or Celestia. Mangan based his action on the radio pioneers who, seizing certain bands or frequencies, "forgot to claim the basic space ... which moves the electronic wave through the sky" (Mangan, 1956). Besides this, according to Mangan, "we all travel through the sky 1,500,000 miles each day [hence] we are basically sky people – not earth people" (Mangan, 1958, p. 7).

On December 29th, 1948, Celestia formally applied for membership in the United Nations. Six weeks later, word came back from the UN Headquarters that Celestia could not meet the provisions of Article 4.1 of the UN Charter, reading:-

> Membership in the United Nations is open to all ... States which accept the obligations contained in the present Charter and, in the judgment of the Organization, are able and willing to carry out these obligations.

Denied a seat at the UN, Mangan became very critical of the organization and of its extraterrestrial interests. In April 1956, he shared his disapproval for the:-

> United Nations, various governments, societies, committees and associations [who] continue to debate and speculate on the future of 'international space law'. They are seemingly unaware of the Celestial fait accompli which makes such deliberations legally impolitic and scientifically impolite (Mangan, 1956).

One year later, he declared Celestia as being "the only political entity to set up the coming space laws that all nations of the earth must obey" (United Press, 1957, p. A4). On June 9th, 1958, Mangan was received by UN General Counsel Oscar Schachter, who was duly and formally told that "no land nation or combination of land nations can write any space laws" (Mangan, 1958, p. 25).

In defiance of Mangan's entity, shortly after the launching of the first artificial satellite, the United Nations General Assembly decided to set up an ad hoc Committee on the Peaceful Uses of Outer Space aimed at considering, inter alia, the legal problems which might result from space exploration. The UNCOPUOS was established as a permanent body in 1959. In May, prior to the UN meeting, Mangan informed General Secretary Dag Hammarskjöld that the summit discussing the future use of celestial space is "out of order and violates the United Nations Charter. All celestial space has for ten years been the sovereign territory of the Nation of Celestial Space." (Letter dated May 5, 1959, Celestia archives).

In the December 21st, 1948 "*Declaration by the Nation of Celestial Space*", Mangan proclaimed that:-

> the use of its space, without permission, is hereby publicly forbidden to any artificial thing, project or activity, not commanded entirely by natural design or need (Mangan, 1956).

Throughout the years, the leader of Celestia would repeatedly attempt to enforce this norm, either by licensing applicants or those whom he considered worthy, or by protesting trespass by those not holding licenses (Mangan, 1958).

One of the most fundamental characteristics of a state is regulation of property rights. Mangan did not claim celestial space for his exclusive benefit; instead, he intended to subdivide his domain and share it with others:-

> so that any man, woman, or child, however small or insignificant they be, may some day own more real estate than the very country he lives in, yea more than the countries of the world combined! (Mangan, 1956).

"No lawyer can deny the claim is valid, because space is real estate"– declared Mangan (United Press, 1949, p. 6A). The "parcels of space" were to be "spheres of pure space, slightly larger than the earth, located by celestial latitude and longitude and other exact means" (Mangan, 1956). Although several people sent in money, the leader of Celestia returned it, seeing no hurry to sell as yet space. It was rumoured that Mangan sold lots on the Moon, yet his son denies this:

> My father knew that one of the international legal requirements for claiming land was that the discoverer must touch the land, something that he himself never had done with the Moon. Then you can legally claim it (Suburbanite Economist, 1970).

Whereas Mangan did not hold the United Nations at high regard, other claimants begged to differ. On February 13, 1952, a Berkley science fiction fan club filed a legal claim for a triangular lunar area containing the craters Ritter, Manners, and Sabine in Mare Tranquilitatis. Letters were sent to the Director of the UN Legal Department, Oscar Schachter and to the US President Harry Truman, informing of the action and asking for comment:-

> Be it known to the United Nations that the Elves', Gnomes' and Little Men's Science Fiction, Chowder and Marching Society, has discovered deposits of sylvanite ... on the Moon and does hereby set up claim to the following area ... as detailed in the attached map. The said society petitions the United Nations to issue to it a patent and title to the above-described area, in consideration of which the said society will assign to the United Nations ninety per cent of all profits that may accrue from the development and sale of the said sylvanite (Fabun, 1952, pp. 4–5).

The news story, along with a print of the claim and the Moon map, was sent to the press. The moon claim was a public relations ploy aimed at boosting the club's reputation in the Bay Area; instead, it made the headlines across the globe. In a 1952 interview, Les Cole, the club chairman, declared having "told the UN ... that the problem of moon ownership will be a problem actually facing the UN or somebody within the next 10 years ... [s]o we don't believe we are presenting the problem prematurely" (Humboldt Standard, 1952, p. 25). Whereas President Harry Truman never replied, it took several letters to Oscar Schachter to get a reply stating that the UN had no jurisdiction and therefore could not do what the "Little Men" asked (Cole, 1996, p. 34).

The UN jurisdiction over the Moon has been the concern of yet another science fiction organization. In the fall of 1952, Chicago hosted the Tenth Anniversary Science Fiction Convention. As a "newsworthy but legitimate means of attracting publicity to the science-fiction field", as stated by the organizers, each one of the 870 participants was assigned a different crater by the organizers "as his exclusive property, to trade, sell, colonize, or ignore, whichever he chooses". The Convention provided for an "official" Lunar Land Commission to manage problems arising from property claims (Reisenberg, 1952, p. 8). "Technically, Earth's satellite is not the Convention's property to bestow" – confessed in January 1952 "First Lunar Commissioner" Mark Reisenberg – "but such slight details have not stopped science-fiction fans in the past from laying claim to various parts of the universe" (Reisenberg, 1952, p. 8). Two months later, when commenting on the Little Men's rival claim with the United Nations for a lunar slice, the organizers declared:-

> In such cases ... the UN has no jurisdiction. The regulation of lunar real estate transactions is not under the authority of terran courts, but remains a speculative problem – a problem of science-fiction. And as the body now representing the largest number of science-fiction enthusiasts, the convention Moon Commission reaffirms its position as final authority in lunar disputes (TASFIC, 1952, p. 8).

Whereas most of the extraterrestrial claimants were located in the United States, the North Americans did not have a monopoly. In 1953, a Chilean lawyer claimed the Moon in order to make a point, his application for membership at a local club having been rejected due to him not owning any real property. The story – quite popular in Chile – says that Jenaro Gajardo Vera went to the office of the local "Conservador de Bienes Raíces" – entity entrusted with regulating the registration of real estate – where he was told he ought to publish his claim three times in the media. As no objections were lodged following the publication, Gajardo returned to Talca's real estate registry where it is told he obtained a document confirming he is the owner of the Moon. (Ricardo Muñoz Cornejo, personal communication, October 10, 2005).

More than a decade later – narrates the legend – the United States had to obtain Gajardo's permission for the Apollo crew to visit what was his private property. It is thus said that President Nixon sent the Chilean lawyer a dispatch reading:

> On behalf of the people of the United States, I request your authorization so that astronauts Aldrin, Collins and Armstrong may land on the lunar satellite that belongs to you (Rojas, 2004).

Obviously, no such letter could have been sent by the US president – if only for the fact that the Apollo landing crews always numbered two astronauts, with the third one remaining in lunar orbit.

Another anecdote has it that the lunar claim caused Gajardo difficulties with the Chilean Bureau of Internal Revenue for not listing the Moon on his real property tax return. Having been visited by fiscal inspectors for the alleged tax evasion, Gajardo invited them to go to the Moon and survey it themselves, and then they can talk about the money he owed. Naturally, the inspectors never bothered him again (El Centro, 2004).

In the year 2000, two years after Gajardo's death, the account took an unexpected turn when two people appeared in Talca, looking for persons sharing his family name. Pretending to be from a law office, they were informing the gullible Gajardos that NASA needed their permission in order to install bases on the Moon, which would bring them important monetary benefits. The two were nonetheless asking for considerable fees in order to regularize the legacy, readily provided by the credulous Chileans. Once they gathered enough money, the con men disappeared (El Centro, 2004). The crooks informed Gajardo's "successors" that in 2005 a law will come into force, authorizing the possession of extraterrestrial goods. As inheritors of a sizeable part of the Moon, each person approached was asked to pay "stamp duties" and other "fees" amounting to the equivalent of 6,000 US$ (Pinto, 2004).

During the next years, people throughout the world would lodge many more extraterrestrial claims. Some of the claimants were business-minded, intending to "sell" their extraterrestrial possessions. In September 1954, three men from Arkansas formed the "Planet Mars Development Corp." with the aim to "subdivide and convey title to such area or areas of said planet Mars to competent persons for suitable remuneration". In submitting the articles of incorporation to the Secretary of State, the businessmen outlined several incentives for prospective buyers of Martian real estate, such as the (supposed) lack of mountains on Mars that would prevent "road-building troubles", while the absence of water would spare any "drainage or irrigation problems" (Associated Press, 1954).

In November 1955, Robert R. Coles, former chairman of Manhattan's Hayden Planetarium, incorporated a business devoted to public education, the exploration of planets, and development of satellites and rockets. The most important activity of the "Interplanetary Development Corporation" (IDC) was, nonetheless, the "sale" of lunar land. For one dollar, the "General Quitclaim Deed to One Acre of Land on the Moon (Northeast Quadrant)" entitled the buyer to an acre of land located in the crater Copernicus (Rhodes, 1955). The purchaser would also receive a postcard that could be returned to the company headquarters requesting a reservation of a seat on their first passenger rocket to the Moon. The plot was identified on the basis of the lot number, imprinted on the cover of deed. The document provided as well for the transfer of "all mineral and oil rights, if any" and "the appurtenances and all the estate and rights of the party of the first part in and to said premises". The IDC ("the party of the first part") awarded the buyer also:

> the privilege of prospecting for all manner of minerals and other elements in the Lunar Apennines, reserving unto the grantor, however, fifty (50) per centum of all minerals and elements of whatsoever nature for the use and benefit of the grantor...

While the quitclaim deed was written in nifty legalese, its provisions were clearly jocular, and Coles was not attempting to deceive prospective buyers. "Anybody who's half intelligent who reads the deed will realize it's a joke", Coles assured a Newsday reporter. "We're not trying to put anything over on anybody" (Rhodes, 1955). And indeed, a dollar provided the buyer with:

> beach rights to, and right of access to, the Sea of Tranquility, for the enjoyment thereof [;] ... the right to fish, dredge and clam in, upon and under the Sea of Nectar [;] ... the right to engage in Winter sports in the Lunar Alps with all the privileges appurtenant thereto without the grantor being in any manner liable for the consequences.

In 1956, the Japan Astronautical Society (JAS) – an entity established by science journalist Mitsuo Harada with the aim of boosting the Japanese public interest in the nascent space age – started parcelling 80-acre plots on the Red Planet. The buyers would receive a JAS membership pin, a guide to planet Mars and a certificate of "Proof of Acceptance of Reservation for Martian Land Allotment", stating one's name, reservation number and the following legalese:

> You possess 100,000 tsubo of land on Mars. In the future when this society succeeds in its plan of developing the soils located on Mars, it will give priority to your reservation and presents this certificate as a guarantee that it will subdivide 100,000 tsubo of land within the planned area (Inoshita, 1957, p. 7).

The Martian land rush spread beyond the boundaries of Japan, with the JAS catering for applications from places as diverse as the United States, New Zealand, Turkey, India, Hong Kong, Argentina and even Africa (International News Service, 1957, p. 1).

The beginning of the space age saw a huge public interest in the extraterrestrial real estate phenomenon. In fact, sometimes the demand exceeded the offer. At the 1956 Auckland Birthday Carnival, the Auckland Astronomical Society had to put up a notice on its booth reading – "We do not sell plots on the Moon", because one out of every ten visitors entered the booth with the intention to buy a lunar property. (Reuter, 1956).

Besides private individuals and clubs, several municipalities in the United States have "officially" annexed the Moon and Venus for fundraising purposes. In July 1965, when the Oklahoma Science and Arts Foundation needed money to pay for a scale model of the Moon, Oklahoma City Council offered a helping hand – it annexed the entire 9 billion acres of the Moon, and Mayor George Shirk turned them over to the Foundation (United Press International, 1965). Inspired by this feat, Lee Bishop, president of the Deer Park Chamber of Commerce, convinced the local council to annex planet Venus, with the intention to sell extraterrestrial lots as a fund raiser. The Minutes of the Regular 415th Council Meeting of the Council of the City of Deer Park, Texas, August 31, 1965 mention that:

> After some discussion, it was moved by Councilman Black and seconded by Councilman Young that we publish our intentions of annexation of the Planet Venus as required by law. The motion passed 7 to 0.

The sale of Venusian deeds helped the city to produce a promotional film and boosted its fame. "People have heard of Deer Park who probably never would have. The publicity, even though it was in jest, helped", boasted Bishop. Yet the Texan city was not the only entity with eyes for the morning star. When in 1967 the USSR became the first nation to successfully send a space probe into the Venusian atmosphere, no request for permission was sought from Deer Park. Bishop was initially tempted to lodge an official protest with the Soviet embassy for having landed on private property and demand compensation; he eventually decided against it, fearful that this would affect international relations (United Press International, 1967).

On April the 6th, 1966, in celebration of a century from the incorporation of the city of Geneva, Ohio, 35 representatives of its citizenry signed a "Declaration of Lunar Ownership". Emulating the US "Declaration of Independence", the document avowed:

> When in the course of human events and space-age accomplishments, the destiny of mankind becomes influenced ... [by] the presence of a particular controversial Celestial Body unclaimed and unregulated ... it should be advisable and honorable ... to lay definitive and prior claim to the entire physical mass and any and all aura, aspect, imaginative or otherwise, of ... the Moon.

The Ohioans consequently assumed "full possession and complete responsibility" of the Moon, bowing "to no man or State in its sovereign right so to do". The document declared Geneva's claim as "positive, supreme, permanent and sovereign", surmounting and voiding "all other claims both real and fancied ... previously made for the possession of all or any part of the Moon". The municipality kindly allowed all humans "to enjoy and to behold and bask" in the moonlight, while refuting liability for "any mental, physical, or spiritual influences ... [or] for any advantages or disadvantages from tidal phenomenon". Space explorers were to be charged per hour for "[l]andings and close fly bys of the visible acreage of the Moon", while requiring "the express consent of the official Councils of the City of Geneva" for any activities on the farside. The document provided as well for the sale, rent or lease of the visible lunar face "to desirable applicants upon a two-thirds vote of the entire population of Geneva", granting the US Government "prior option to this privilege". For the year 1966, the document allowed the sale of 100 deeds "for the sum of $100 describing 100 acres from Mare Umbriam"; "[t]he possessor of this deed shall have it to hold, his heirs and assigns forever and unencumbered".

The "extraterrestrial estate" agency that received the broadest publicity is the Lunar Embassy, founded in 1980 by a Californian, Dennis Hope. Unaware of the previous lunar real estate affairs, and convinced that the 1967 Outer Space Treaty prohibition of national appropriation in outer space would not apply to individuals, Hope rushed to the San Francisco County offices where, on November 22nd, 1980, he registered a "Declaration of Ownership" notifying the "sovereign planet of, Earth" that he is now and shall ever be known as "the omnipitant ruler of the lighted lunar surface", with "the exalted title of, 'The Head Cheese'". The odd-spelled document informed the world of Hope's claim for the ownership of Earth's Moon and of the remaining known eight planets and their respective moons: "As owner I hold all rights over these properties without limitation". Thus, the declaration stated that transactions in regard to real estate planning, development, and further exploration of the illuminated lunar surface are to be done subject to his advice and consent. Also claimed were "[a]ll existing mineral, water, oil, and liquid rights", the claims being made as his "forefathers claimed the property known to all mankind as, earth". Hope declared "all known rules and laws pertaining to homestead acts as known on the planet, Earth" as being "invalid on the above listed planets and moons", with him having "total governmental say as to tenants and governmental entities that wish to negotiate any involvement with the Lighted Lunar Surface."

The next step was to apply for a copyright and to form and register his company, "The Sovereign Worlds of Hope", a.k.a. "The Lunar Embassy", with the Contra Costa County offices. Copies of the "Declaration of Ownership" were sent to the governments of the US, USSR, and to the UN General Assembly, together with a note stating his intention to sell the claimed property. The governments were sent also a bill for $55,000 for storage and littering on his property (Hope, 1997).

Hope notified the USA and USSR "as a courtesy only", being "the world powers at the time" while he regarded the United Nations as "the only organization on this planet that was recognized as having the authority to create laws for deep space". Indeed, Hope values the United Nations as "the source" of his claims:

> In their own language, 'No nation by appropriation shall have sovereignty or control over any of the satellite bodies'. Registering property on other planets should have been done with the recognized authority for such claims. Since the language is specific in the Outer Space Treaty of 1967 from the UN it is safe to conclude they did not own the land either. As the representative group for the creation of laws in space, our legal advisors have stated our claim is legal and binding (Dennis Hope, personal communication, December 1, 2005).

The first sixteen years of the Lunar Embassy were unremarkable, with only 3,500 "properties" sold. Since 1996, when the first facility to house the business was built, the venture took off. Dennis Hope has subsequently demonstrated mastery in using several strategies, especially in harnessing the power of the Internet since 1998. Another avenue to increase sales was the start of a two-level, fee-based reselling program. The "Authorised Reseller" tier was conceived for those individuals willing to sell extraterrestrial properties in their country on a part-time basis. In December 2005, there were 27 reselling agents. For those eager for deeper involvement as exclusive resellers for an entire country, Hope created the "Ambassadorship" program, currently represented in 15 countries.

The large number of clients attracted by the Lunar Embassy and the accompanying media frenzy, coupled with the "why didn't I think of that" syndrome, resulted in other entrepreneurs imitating Dennis Hope. Copy cat companies are seen by Hope as frauds, given that they "did not file the claim of ownership at the appropriate time". Whereas the first 8 or 9 companies were told to cease and desist and they did, suing the remaining copycats is not economically sound. At the cost of more than US$ 40,000 for a major lawsuit, this would be wasted money, says Hope, as "you could sue them here in the USA and win. Then two weeks later the copy cat would start a new business under a new name and create the same problem for you" (Dennis Hope, personal communication, December 1, 2005).

And the story could go on and on. The success of the extraterrestrial real estate companies – willing to cater for the ancestral yearning of holding the moon in one's hands – is understandable. Even if they do not actually convey property rights on Earth's natural satellite, their proliferation shows that there is a need for extraterrestrial property rights. Besides this, the "land-on-the-moon-for-sale" stories have the required "wow!" factor to entice the media – an often-unwitting ally of the "unreal estate" outfits through the free publicity provided. The press is rarely interested in genuine issues of space law, and when articles appear, they deal instead with the Lunar Embassy and the like. The media has reported throughout the years on many instances of "sales" of alien plots, but as the newspaper pages yellowed, the stories went forgotten. Not in the case of Dennis Hope – with the advent of the Internet, the Lunar Embassy has emerged as the wrong institution at the right time in the right place.

Some reports, ignorant of the real legal provisions applicable to the extraterrestrial realms, feature the whole business on a positive tone, while most of them use a jocular one. Many newswriters celebrate Mr. Hope's ingenuity at spotting loopholes without bothering to check the legality of the situation while – true enough – another part of the media has sought the opinion of legal experts. In an era of moral relativism, where statements are taken at face value, readers have commonly been left to judge for themselves who is right: the entrepreneur, or the lawyer. This unpleasant situation needs to be challenged.

1.3 Arguments For Invalidating the "Extraterrestrial Real Estate" Claims

In terms of space law the purported sale of extraterrestrial real estate is a trivial issue, pertaining, *ratione materiae*, to consumer law. Space lawyers may agree to disagree on some matters, but there is a consensus in the community that no individual or company currently owns the Moon, hence nobody may sell lunar land. While a specialist quickly dismisses the matter as not worthy of investigation, there is nonetheless a world outside the ivory tower of the profession, a world not knowledgeable of space law issues.

Trivial as may be for a space lawyer, it is not so for the lay people. The popularity of the Lunar Embassy has hijacked the public perception of the space law. In the public mind, space law does not mean return and rescue, it does not mean registration, nor does it mean regulation of remote sensing. For the regular person on the street, space law concerns the sale of extraterrestrial real estate.

Yet, some members of the profession have understood that, in order to constructively discuss the issue of property rights in outer space, they have to pay however attention to this trivial issue. If one wants to build on a plot of land, one has to demolish the derelict house standing in the middle of the plot. This needs to be done with legal arguments, an outline of which it follows.

Due to its particular importance in the public eye, and for the convenience of analysis, Dennis Hope and his Lunar Embassy are scrutinized here as the most prominent example, yet the analysis applies, *mutatis mutandis*, to other self professed extraterrestrial realtors.

1.3.1 The Non-Appropriation Principle

In 1938, columnist Raymond P. Coffman received a letter from A. D. Lindsay, in which the landlord of the "Archapellago" boasted about his celestial possessions. Unlike other members of the media, Coffman was not fond of Mr. Lindsay's extraterrestrial dominion: "I have taken a good look at the star-filled sky, and have wondered how anyone could claim to own it" – replied the columnist, carrying on:

> Could Irwin County ever have the right to give deeds to the heavenly bodies? Does any county own them, or does the United States own them? Could all the nations of the earth put together be said to own them or to have the right to dispose of them?

Likewise, Coffman became the advocate of the inhabitants of other planets: "How about the claims of people on Mars or any other planet (if there are such people) to the Sun?" The solar system of the 1930s was a luxurious realm, populated with green Martians and amazons from Venus. And his defence reached as far as Lindsay's imagination: "How about the claims to the stars of people of millions of planets which may be spread through our universe?" Questions pondered years later by the pioneers of space law were raised in a children's column in 1938. Even more, Coffman's conclusion was a statement ahead of time of what the UN treaties will proclaim many decades later: "More important" – said he – "I would not want to 'own' the Sun and Moon and the stars. They seem to me the common property of all who can enjoy them." (Coffman, 1938, p. 12)

The first reason for invalidating the many extraterrestrial claims presented *supra* is the non-appropriation principle of space law. Article 11.3 of the Moon Agreement of 1979 prohibits *expressis verbis* the appropriation of "the surface ... [and] the subsurface of the Moon" by:

> any State, international intergovernmental or non-governmental organisation, national organisation or non-governmental entity or of any natural person.

Nevertheless, the Moon Agreement has only received a small number of ratifications and, by the fact of the United States not being a party to the Agreement, private American claimants are well entitled to say that its provisions do not apply to them.

It is notable, however, that the United States and a significant number of other countries have ratified the 1967 Outer Space Treaty, that in Article II forbids the:

> national appropriation by claim of sovereignty, by means of use or occupation, or by any other means.

of outer space, including the Moon and other celestial bodies.

Dennis Hope maintains that this norm contains a loophole allowing the private appropriation of the extraterrestrial realms. In his words, while the Outer Space Treaty "explicitly forbids any government from claiming . . . the Moon or a planet", what is important "is what the Outer Space Treaty does not say. It explicitly does not say whether commercial enterprises or private individuals can claim, exploit or appropriate the celestial bodies for profit" (Lunar Embassy, 2006b, para. 1). As the Outer Space Treaty has somehow "neglected to mention individuals or corporations", the UN "attempted to plug this loophole . . . by introducing the ill-fated Moon Treaty" (Lunar Embassy, 2006a, para. 2).

As detailed *infra*, this is flawed logic. Thus, if States cannot appropriate the extraterrestrial realms, then *a fortiori* neither can their nationals. Furthermore, landed property rights are unlikely to survive without protection from a sovereign entity, and State endorsement of private appropriation would be a form of national appropriation. A number of academics consider that not only active endorsement, but even a passive attitude from the authorities is a form of national appropriation; as detailed by the Dutch space lawyer Frans Von der Dunk (2000), during a discussion on the issue that took place in 1999 at the 42nd Colloquium on the Law of Outer Space, the majority of the lawyers present there were of the opinion that:

> the United States would actually be in violation of the pertinent rules of the Outer Space Treaty (i.e. that national sovereignty on a territorial basis does not apply to outer space) by not taking action against its citizens trying to offer pieces of the moon for sale, which was amounting to fraud.

While the Moon may have been claimable in the past – being free from the retroactive application of the contemporary non-appropriation principle –, even then a *corpus* was necessary besides *animus*, as detailed next.

1.3.2 Claiming Does Not Mean Owning

The second reason for invalidating the claims presented above is the lack of *corpus possidendi*. In plain language, claiming does not mean owning. In the acquisition of possession, two concurrent elements – "the mind" and "the body" are required. One is insufficient without another; there must be "both an intention to take the thing and some act of a physical nature giving effect to that intention" (Reid, 1996, p. 103).

The first element required is the "*animus possidendi*", the intention to possess. The fact that lots on the Moon were peddled as far back as the 19th Century – and maybe even earlier – is not surprising. Nor is the fact that an impressive number of people availed themselves on the services of lunar realtors. The desire to own the

Moon, even a tiny slice, is hardwired into the human mind. There is no child who has not raised, at least once, a hand towards the Moon and stars, trying to grab them. Thus, it is narrated that Alexander the Great wept when hearing from Anaxarchus that there are countless orbs in the universe: "Do you not think it a matter worthy of lamentation" – moaned he – "that when there is such a vast multitude of worlds, we have not yet conquered one?" (Plutarchus, c. 46–120, IV). His words were echoed, centuries later, by Cecil Rhodes, who lamented in the year of his death:

> The world is nearly all parcelled out, and what there is left of it is being divided up, conquered, and colonized. To think of these stars that you see overhead at night, these vast worlds which we can never reach. I would annex the planets if I could; I often think of that. It makes me sad to see them so clear and yet so far (Rhodes, 1902).

Greedy as they were, neither Alexander, nor Rhodes added the celestial objects of their craving to their maps. They knew that thought alone does not conquer the Moon and the stars, nor does simple proclamation. Yet there were many others who, with differing degrees of conviction, thought they were the first ones to embed the flag of their desire on the alien orbs.

Lunar salesmen can not own the Moon just because they want to. They lack the second element required in the acquisition of possession, namely the "*corpus possidendi*"; without an act of physical nature giving effect to the intention to take the thing, animus is insufficient. The Scottish jurist Stair (1693, II.i.18) has explained this in very illustrative terms: "if any act of the mind were enough, possession would be very large and but imaginary". An imagination as large as the Universe, in the case of the purported extraterrestrial salesmen, who present a very valid *animus*, but no *corpus* at all. By *reductio ad absurdum*, should this not be the case, it would be impossible to prove that there is nobody else that merely wished to own the Moon and the stars beforehand.

Our argument is shared by Philip McDougal (quoted in Naysmith, (2001)) associated legal officer at the UN Office for Outer Space Affairs; in the matter of the Lunar Embassy, he declared that "[n]o legal system allows for the claim of ownership rights in this way." Nevertheless, the legal system of the Masai tribe in Eastern Africa works on the premise that, by the divine command of their God N'gai, all the cattle in the world belong to them (Florida Geographic Alliance, 1992), even though some may have temporarily found themselves in the possession of others. When the Masai raid their neighbours and seize their cows, they are vindicated by their legal system: they are simply returning the cattle to their rightful owners (Stumpf, 1998). If a claim alone would entail ownership, this would entitle the Masai to universal cattle dominion. Yet, in reality, people continue to buy and sell cows without involving the Masai, regardless of what the latter believe. The "it is mine, because I say so" approach – be it in its bovine or lunar form – is invalid.

A number of would-be extraterrestrial landlords made use of county recorders, particularly in the United States, with the belief that entering a claim would make them owners. Many of them were unaware that recording a claim is a mere means of informing the public of the claim and of setting a specific date, and not an act securing ownership. According to Barbara Frerichs (personal communication, November

25, 2005), Administration Manager at the Maricopa County Recorder – repository of several extraterrestrial claims –, the mere fact of recording a document with their office does not make it legal *per se*. If the document is not legal *ab initio*, the act of recording it will not change that. The Recorder office does not decide legality, this being the role of the court if the document were challenged. Nor is its role to verify whether a piece of property is owned or claimed by some else. As to the effect of recording a document, it simply makes that document a public record for notice to the public. Spokane County Auditor Vicky M Dalton (personal communication, September 24, 2005), where one Thomas P. Budnick recorded many extraterrestrial claims, concurs; the most significant benefits of recording any document with the Recorder are to put the public on notice and to set a specific date that the document was recorded. The said institution does not determine the legality or validity of these documents, being, in Dalton's words, "simply a library open to the public."

Well aware of Jenaro Gajardo's story and the numerous assumptions surrounding it, Ricardo Muñoz Cornejo of Talca's Conservador de Bienes Raíces is happy to shed light on the legal implications of recording the Moon with his institution. According to the civil servant, there is actually no official record of the said inscription in the registries of the city of Talca. Apparently, the Recorder office at that time would not register the Moon, objecting that its boundaries were not precise. The certificate in Gajardo's possession, presumes Muñoz, may have actually been an affidavit, a document that is not sufficient to make Gajardo the owner of the Moon. "Everything else pertains to the legend that has formed around this unique character, remembered by the historians also as a great joker" (Ricardo Muñoz Cornejo, personal communication, October 10, 2005).

While a number of would-be extraterrestrial landlords managed to file claims with county recorders, some clerks rejected extraterrestrial claims as not falling under their jurisdiction. Since 1997, the Archimedes Institute – an aerospace policy research organization – has been maintaining a claim registration office, aimed precisely at recording extraterrestrial claims. By the end of 2005, almost four hundred claims have been recorded online. The initiators of the Archimedes Institute registry motivated their decision by the legal uncertainties in the space investment equation. One of these risks is the absence of a property regime for space resources. The organization considers that, while the best way of eliminating such uncertainties is the creation of an international agreement regulating the claim and transferral of space property, there is currently no political will or public interest in creating such a regime. Therefore, they envisaged a private system of property registration at the international level. While the register is not protected by international or municipal law, it would benefit both the registrant and the space development community as a whole. The registry is seen by its initiators as a means of enabling an interested party – be it an individual, corporation or another entity – to claim to the public his intent to retain control or develop a site, thus discouraging a competitor from investing in the development of the same area. The registry is also to be of benefit in court challenges to the said resources, having established "unbiased and unquestionable evidence of the intent and activity of the claimant". On a larger scale, the registry is deemed to encourage the establishment of a "new, efficient and equitable" legal standard for extraterrestrial property claims registration, providing "both an intellectual focus and the political impetus for a formal international registry financed by the nations of the world" (The Archimedes Institute, 1997a). Like its terrestrial counterparts, the Archimedes Institute does not confer ownership, and is unambiguous about this, clearly stating in its disclaimer:

> The Institute makes no warranties either express or implied regarding the validity of the claims filed with the registries by any and all claimants . . . The Archimedes Institute makes no claims of ownership on space resources by virtue of this Registry. While the Registry may be useful in evidencing ancient claims, it does not affect the validity (or lack thereof) of any claims made prior to its establishment (The Archimedes Institute, 1997b).

The lack of *corpus possidendi* may however change in Dennis Hope's instance; while still maintaining to already own the Moon and other celestial bodies, Hope publicized his intention to establish a *corpus*. In March 2001, he initiated "the Affidavit Project", calling the willing Lunar Embassy clients to write an Affidavit, describing their lunar property, on cigarette paper. Hope plans to gather about 1 million such Affidavits that he intends to send to a designated area on the Moon:

> The purpose of the project is to physically represent each of the participating property owners to place a signed document on the surface of the Moon in order to strengthen, through physical possession, your ownership of the purchased Lunar property (Lunar Embassy, 2001).

Mr. Hope plans to send his Registry Archive to the Moon with what appears to be TransOrbital's *Trailblazer* mission. Nonetheless, the *Trailblazer* is open to everybody, not only to the Lunar Embassy, and in exchange for $16.95 (TransOrbital, 2002) one can send one's own concurrent claim to the Moon.

Should Hope succeed to send his archive on the Moon, then – and only then – may he show evidence of *corpus* in support of his claim, though at that time further legal problems will need to be discussed. From a legal point of view, sending one's claim to the Moon would no doubt enhance it, but, at the very best, its effect would be limited to a pint-sized lunar plot. And it remains the question of compatibility with the Outer Space Treaty's non-appropriation principle, together with the fact that Mr. Hope claims to own more celestial bodies than the soon-to-be-impacted Moon. The latest argument is dismissed by Mr. Hope (personal communication, December 12, 2005): "The Moon of Earth is the headquarters for the Galactic Government. We have designated occupation of the Moon as representing occupation for all our territories."

1.3.3 Prior Claims

According to the Lunar Embassy (2006a, para. 2) website, "one can become the legal owner of an extraterrestrial body, if you are the first one that claimed it, and that is the Lunar Embassy". While the first part of the statement is legally inaccurate, the second thesis asserting priority is factually inaccurate. The Lunar Embassy is not the first entity to have claimed the Moon. If wolves and dogs have howled at it since time immemorial, many actual claims precede Dennis Hope's one, dating from 1980 – as seen *supra*. "Prior to my claim in 1980 there were, as I have found out much later, other claims in other parts of the world", Mr. Hope agreed in 2005. As to the validity of the deeds issued by extraterrestrial real estate companies before the Lunar Embassy came into being, he dismisses the claims predating the Outer Space Treaty as being "null and void":

> Any claims existing prior to the Outer Space Treaty have no basis in reality because the filings were merely done on a local basis with governments that did not have control over those claims. Since the UN put itself as the authority for laws in space they become the recognized source (Dennis Hope, personal communication, December 12, 2005).

Even this is flawed logic, as his claim is not the first one advanced to the United Nations, and not the first claim following the enactment of the Outer Space Treaty. The most logical question one could ask would be – if Mr. Hope's is just one of the many claims, and not even the first one, why do all the people "buy" lunar plots from him, and not from previous claimants? The answer has nothing to do with the alleged lawfulness of Mr. Hope's claim, but with the advent of the Internet and with Mr. Hope's marketing genius in taking advantage of this new means of communication. His claim was made back in 1980, but the business did not develop until the mid 1990s, when his website came online. In contrast, the old claimants did not have the Internet at that time, so their ventures never really took off.

1.3.4 Qui Tacet Negat: Silence of Authorities is Not Acquiescence

In sustaining his claims, Dennis Hope also invokes the silence of the authorities, both US and foreign. The Lunar Embassy (2006a, para. 2) felt "obliged to inform the General Assembly of the United Nations, and the Russian Government in writing of the claim and the legal intent of selling extraterrestrial properties". However, "[t]he US Government has several years to contest such a claim. They never did. Neither did the United Nations nor the Russian Government."

International law requires, for the establishment of an historic title, that the claimant exercise authority that is both continuous and peaceful. The first requirement is met by exercising effective possession over the disputed territory, while the second by the lack of opposition to such display of authority. Regarding the first element, exercise of possession must be done *a titre de souverain* (Blum, 1965, pp. 99–100). But such possession *a titre de souverain* of the Moon would be unlawful, as it would violate the non-appropriation principle of the Outer Space Treaty. Furthermore, as seen *supra*, the above mentioned claimants do not have any kind of possession over the extraterrestrial realms, lacking the *corpus*. Regarding the second element, its presence would matter should possession a *titre de souverain* be met. However, not even the second element exists in Hope's case. As Blum (1965, pp. 130–131) notes:

> the absence of protest is relevant in the formation of an historic title only in those cases in which protest would have been expected to be forthcoming, had the affected State

> really wished its objection to be made known. There are situations ... in which an inference of acquiescence cannot be justifiably drawn from the simple fact of absence of protest.

We believe that protest was not to be expected from the UN and USSR when confronted with such trivial claim; in this case, the maxim *qui tacet consentire videtur* is to be read as *qui tacet negat*. In fact, the Soviet reaction to earlier lunar real estate affairs was that of a good laugh: "As for appropriating celestial bodies, only American speculators trade in lots on the Moon" – Soviet jurists commented decades ago (Lay and Taubenfeld, 1970, p. 78). As the authorities did not consider the matter worthy of challenge, the absence of objections from the authorities is not relevant. God himself does not send fire and brimstone to all the false prophets who declare themselves as anointed by Him. The same view is hold by Philip McDougall, associated legal officer at the UN Office for Outer Space Affairs. Besides stating that the UN "do not have any record of receiving this correspondence" (Naysmith, 2001), McDougal explains the silence of the UN in very colourful terms:

> I could tell everyone I own the Crown Jewels, set up a website on it and write to the Queen telling her they were no longer her property. Just because she doesn't write back doesn't mean she is acquiescing to it.

The same UN legal officer explains that the reason that nobody has blocked Hope's claims is that he is seen as a harmless eccentric (Ahuja, 2000).

While claiming a celestial body *per se* is not an illegal activity, selling it to the public may be a different matter, depending on the jurisdiction. Authorities have most of the time turned a blind eye to the extraterrestrial realtors, but there have been exceptions to this lenient approach.

On November 22nd, 1955 – a week after Robert R. Coles started selling lunar quitclaim deeds – New York State Attorney General's office launched an investigation aimed at determining whether anyone was being taken advantage of. In announcing the probe, First Assistant Attorney General John Trubin declared that it is the "office's responsibility to protect 'suckers'" (Central Press Canadian, 1955) and that "[d]espite some of the ludicrous aspects of this case, if people are being victimized, even for small sums [i.e., US$1], that is not funny" (Associated Press, 1955). "Anybody who can come up with a clam out of a sea with no water in it will have to be pretty good" – declared Coles in reply, assuring the press he was not attempting to deceive prospective buyers. "Anybody who's half intelligent who reads the deed will realize it's a joke" (Rhodes, 1955). And the letters sent by the buyers inferred that about 90 percent of them were aware of the gag (Steinberg, 1996, p. 168) leaving however a minority of 10 percent who did buy the deeds in earnest.

In August 1969, Brazilian police in Belo Horizonte arrested one Jose Cassiano De Jesus for selling lunar lots. As a marketing ploy, he was telling prospective clients that his first customer was Neil Armstrong, and that the astronaut had gone to the Moon in order to inspect his property (Associated Press, 1969). The price for a plot on the near side of the Moon was $145. While the customers were embarrassed, none of them wanted to press charges, hence De Jesus was released (United Press International, 1969).

In May 1975, a federal grand jury in Tucson, Arizona, indicted Arnold O. Morales on 19 counts of mail fraud, for allegedly advertising the offer for the sale of deeds to 1,000-acre parcels on the Moon for $4.98 in a national magazine. Morales received about 900 replies the week after the advert ran; according to US postal service officials, some persons enclosed money along with their replies (Associated Press, 1975). The advertisement did however specify that the gift was a "document that almost looks official and legal", being a "novelty item only – not a legal instrument". The order form also contained the statement – "I also understand that this is only a novelty item and in no way is to be construed as a legal document" (Moon Acres, 1974, p. 16).

In 1982, the Fiske Planetarium and Science Center launched the "Great Martian Land Sale", a fundraising campaign. When director David A. Aguilar sent a complimentary Martian Deed to US President Ronald Reagan, the White House replied with a dry letter informing Aguilar that he was in violation of international space law, and that a request for investigation had been forwarded to NASA. A gift package then left the Fiske premises, addressed to the NASA administrator. Aguilar received a short note back from NASA Administrator James Beggs, thanking him for his otherworldly gift. (David A. Aguilar, personal communication, October 17, 2005).

As seen *supra*, the practice of the authorities in the matter of extraterrestrial real estate is far from homogenous. Aware of the need for a stricter approach, the Board of Directors of the International Institute of Space Law issued a Statement in 2004 where it deplored the augmentation in the extraterrestrial real estate business "raising the opportunity for individuals to be misled". The statement reads, inter alia:

> The prohibition of national appropriation [of outer space and celestial bodies] ... precludes the application of any national legislation on a territorial basis to validate a 'private claim'. Hence, it is not sufficient for sellers of lunar deeds to point to national law, or the silence of national authorities, to justify their ostensible claims. The sellers of such deeds are unable to acquire legal title to their claims. Accordingly, the deeds they sell have no legal value or significance, and convey no recognized rights whatsoever.

The Board of Directors of the IISL calls State Parties to the Outer Space Treaty to:

> comply with their obligations under ... the Outer Space Treaty ... [being] under a duty to ensure that, in their legal systems, transactions regarding claims to property rights to the Moon and other celestial bodies or parts thereof, have no legal significance or recognised legal effect (International Institute of Space Law, 2004, p. 2).

It is hereby suggested that the IISL disseminates this Statement as broadly as possible, particularly to consumer protection entities. Perhaps, it would be indicated to be joined by the United Nations Office for Outer Space Affairs when doing so. This way, it is hoped the national authorities will react more vigorously to this annoying business.

1.3.5 *Jocandi Causa*

While Saint-Germain (1997) considers Hope's activity as being "not a joke, and it's perfectly legal", in reality a choice has to be made: it is either a joke, and thus lawful, or it is not a joke, and thus illegal.

In the case of the Lunar Embassy, as it points out itself, the Lunar Deeds are "novel gifts" (Lunar Embassy, 2006a, para. 3). Their legal classification as novelty items means that these are to be used *animus jocandi*, i.e. for fun only. It is not illegal to sell or to possess novelty items; it is illegal though to misuse them outwith the "novelty use only" scope. Other companies sell items such as one-million-dollar bills, or camouflage passports from inexistent countries, or "Area 51" license plates that, as long as they are commercialized and used as "novelty gifts", do not upset the authorities.

Nonetheless, Dennis Hope considers the documents issued by the Lunar Embassy as a combination of quit claim deeds and warranty deeds. In 1980, when the Lunar Embassy was formed, Dennis Hope was advised by legal staff to place the term "novelty gifts" on the deeds. Accompanying this was a definition of the term "novelty" as presented by the Heritage Dictionary, which depicted novelty as a small mass produced item, and something having novel characteristics. "All true in this case. We changed that configuration to 'Novel Gift' in 2000" – states Hope (personal communication, December 12, 2005).

"If I were in court under oath I would categorically refer to myself as the legal owner of the Moon of Earth and the other eight planets and their moons", declared the Lunar Embassy CEO. "We actually sell land not certificates. Since we are the originators of the land sales we are the initial grantor of the titles to these lands and we stand behind our issuance of these deeds". It is not only he who considers that he is selling the real thing, but also the buyers. According to the Lunar Embassy records, 42% of all purchases are assigned to some form of a family or business trust, while only around 17% are believed to be purchased for the fun aspect of the property sales. (Dennis Hope, personal communication, December 12, 2005)

The history of the lunar claims, as outlined *supra*, shows that many were publicity stunts. Such was the case with the "Little Men's" claim, and with Geneva, Ohio's "Declaration of Lunar Ownership" which, capitalizing on the nation's fascination with lunar exploration, was aimed at drawing attention to the city's centennial celebrations (Feather, 2004).

A lawful approach has been taken by a web site operated by Digital Minds, Inc. Considering that paying money for a worthless certificate from the "Lunar Embassy" is a ridiculous practice, Russ Wylie decided to offer one that had some value – albeit not material. Toying with the indelicate name of the seventh planet, Wylie invited the public to "grab a piece of Uranus before someone else gets a hold of it". The potential buyer is motivated by an existential question: "Can you afford not to own Uranus?" – and by the scatological humour of the founders of R.E.C.T.U.M. (Real Estate Commission & Trust of Uranus Management), the registry for the "Uranus Deed Registration". Unlike Dennis Hope's Lunar Embassy, Digital Minds know the law; in the website disclaimer, they acknowledge that:

> BuyUranus.com is a novelty site operated for entertainment purposes only. All of the merchandise for sale on this site is only for entertainment and holds no value other than its entertaining nature. Digital Minds in no way claims that when you buy a piece of Uranus that you have any legal claim to that little piece of the Solar System unless you go up there and put a fence around it personally (Buyuranus.com, 2000).

Their approach is not based on deception, but on pure sense of humor, and the buyer knows that he or she would simply buy a piece of paper with no other value than a good laugh. This is instead of being conned into believing that he or she actually owns that extraterrestrial property, like the "Lunar Embassy" does when stating that a "novel gift":

> does not diminish the value of the property that you purchase in any way, as every deed is recorded and registered in the Lunar Embassy's registration database and every owners information is listed with that registration. You own this property (Lunar Embassy, 2006a, para. 3).

While avoiding lawsuits, being a novelty gift does in fact affect the value of the property, unlike claimed above. In 1997, NASA's news chief Brian Welch declared that he knows of no plans to take legal action in the extraterrestrial real estate affair, as the deeds to lunar property are worthless: "That's why they invented the phrase *Caveat Emptor*" – let the buyer beware (CNN, 1997). The astronaut Buzz Aldrin is also aware of their real value: "Well, if somebody wants to have a certificate that says they own a certain portion of the Moon, and they're willing to pay whatever it is, probably the only thing they'll ever get is a certificate" (Knapp, 1996). Von der Dunk (2000), from the position of the space lawyer, asserts that "you would buy, in other words . . . basically a nice piece of paper; . . . a hollow claim at best."

Should the Lunar Embassy adopt the Buyuranus.com method, and stop conning people into believing it actually owns the Moon and the planets, it would still remain the leader of the "extraterrestrial real estate" industry given the already existing publicity, and people would still buy funny pieces of paper for a laugh, this time in full knowledge of the facts.

1.4 Conclusion

The whole extraterrestrial real estate affair has no support in space law, its relevance pertaining more to the discipline of consumer law. Nonetheless, it needs to be addressed in the quest for the answer to the question "Who owns the Moon". As "*nemo dat quod non habet*", the lunar wills and deeds do not have the legal effect of endowing people with extraterrestrial property. They need not to be formally declared void; they are already so. Should a contract be made in fraud of third persons, it is void *ab initio*; "[n]o person's rights can be affected by it, whether he be a party or a stranger" (Sweet,1882, p. 869). Local councils do not sue "Monopoly" players for "buying" and "owning" High Street – as long as the "board game owners" do not squat the real-world High Street. As shown above, the "lunar deeds" are to be seen as what they really are: jocandi causa "board game certificates". As such, they cannot serve as evidence in real world trials, where the alleged owners of lunar estate and minerals would try defending their "properties" against "trespass" from prospectors and developers.

While the "extraterrestrial real estate" claims described *supra* are nothing more than media curiosities, it needs to be agreed that behind their triviality they hid significant legal implications. The advancement of such claims has been only possible because of the lack of a clear property rights regime in the extraterrestrial realms. In the following chapters, the issues concerning such a regulation will be thoroughly addressed.

Chapter 2
The Sources of Landed Property Rights in Outer Space

A lawyer is not a person who knows the law, but who knows where to find it

King George III

Property and law are born together, and die together. Before laws were made there was not property; take away laws, and property ceases.

Jeremy Bentham (1802, Chapter VIII)

2.1 Introduction

At a first sight, the extraterrestrial realms seem the scene not only of a physical void, but also of a political and legal vacuum in the subject of landed property rights (Pace, 1998; O'Donnell, 1999, p. 2). The Lunar Embassy (2006b, para. 3) puts this in hilarious terms: "On the Moon itself or on any other planets apart from Earth, apart from the laws of the 'Head Cheese', currently no law exists".

Whereas the "laws of the Head Cheese" have been dismantled in the previous chapter, the apparent silence of the law in the field of extraterrestrial landed property rights has been addressed by several commentators. The reasons advanced vary from the absence of an actual need for regulating extraterrestrial property to the inability of the legislator to cope with such a challenge. The question to be asked here is whether the factual situation on the extraterrestrial realms calls for legal regulations. It is to be noted that the "sources of law", including that of extraterrestrial property rights, have a twofold meaning – formal and material. The formal (or normative) aspect of the sources of law is the form in which legal norms manifest themselves – that is, the rules of law as stated in regulations. In their positive sense of *lex lata*, the sources of landed property law in outer space imply the place where the legal norms regulating the regime of extraterrestrial immovables are found. The other meaning illustrates the actual facts that have caused the establishment of these norms, as expressed in the maxim *ex facto sequitur lex* or *ex facto oritur jus*, i.e. law has its origins from the facts. The material (or causative/formative) sources are constituted by the external realty that causes and influences the enactment of law – that is, the technological, political, economic, cultural, social, financial, and other causes that stand at the basis for creation of legal rules and institutions. The material sources concern the need for a particular piece of legislation, these being often referred to in

V. Pop, *Who Owns the Moon?,* Space Regulations Library Series 4,
DOI 10.1007/978-1-4020-9135-3_2, © Springer Science+Business Media B.V. 2009

its preamble; they reveal why a particular norm exists, or does not exist – why law is silent. While the material sources of law consider the challenges and how does one respond to them, the formal sources of law contain the answers to these questions.

Does the law of extraterrestrial landed property conform to the rule *ex facto sequitur lex*? What are the facts that have caused and shaped the law of extraterrestrial real estate? What norms express this law? The following chapter attempts to addresses these questions.

2.2 Real Property Rights Implications of Space Activities

In commenting on the matter of extraterrestrial real estate, NASA's news chief Brian Welch recognized that the otherwise trivial issue could get more serious in the future:

> when people actually are going to these places and the resources found have some value. More complicated issues will have to be resolved between countries or between companies (CNN, 1997).

Indeed, the extraterrestrial realms are not only, as Apollo 8 astronaut Frank Borman views it – "a vast, lonely, forbidding type of existence or expanse of nothing" or, in the words of Apollo 11 astronaut Buzz Aldrin, "magnificent desolation" (Chaikin, 1994, pp. 121 and 211). According to the Science Advisor to the US President and Director of the Office of Science and Technology Policy John Marburger (2006), "[t]he greatest value of the Moon lies neither in science nor in exploration, but in its material" – anything that can be made from lunar resources at a cost compared to terrestrial manufacture having a tremendous cost advantage over objects launched from Earth. Whereas the Apollo program turned out to be a flag-and-footprints endeavour, Marburger envisages a lunar industry dealing with the extraction of minerals and elements and their processing into "fuel or massive components of space apparatus", in a future when the Solar System and its materials are incorporated in the terrestrial economic sphere and way of life. Mark Sonter (1998) too deems that, in order to overcome the high cost of launch from earth, the large scale commercial activities in space will require raw materials obtained from space. Volatiles for manufacturing propellants and metals for construction in space appear to be readily recoverable from asteroids and dormant cometary bodies. The space materials, considers Declan O'Donnell (1998a, p. 29) will be used for building habitats, transportation systems, hotels, factories and so on.

Outer space is abundant in such resources; according to John S. Lewis (1996, pp. 195–196), the asteroids, comets and planets contain untold riches. His calculations show that using asteroidal iron and steel would generate wealth amounting to $7 billion per person; adding in this equation the other ingredients composing the asteroidal belt – such as gold, silver, uranium, etc. – the total would rise to over $100 billion for each person on Earth. But – should humans be billionaires in asteroidal metals –, how is this wealth going to be appropriated and shared? Who, after all, really owns the Moon and the other celestial bodies? What is the place and significance of property rights in the context of space activities?

The existing space activities, as well as their further development, have fundamental property implications; their dimension spans both the terrestrial and the extraterrestrial realms. Important questions are raised by many endeavours – from the stratospheric platforms interaction with the subjacent land to the property status of space debris; from the intellectual property status of inventions, industrial processes and new materials created in the extraterrestrial realms to the ownership of data obtained through remote sensing. The limited dimensions of this book do not allow however a thorough analysis of all these aspects. The research will focus therefore only on the legal situation of the landed extensions, by examining the property status of the celestial bodies *per se*, with reference to the legal nature of the planetary surfaces, of the natural resources thereon, and of the "legal interaction" between planetary surfaces and facilities placed thereon. This field of study is very fertile, the question of "real property rights in outer space" or "planetary land law" posing intriguing problems. One such issue is jurisdictional, and it concerns the search for an appropriate *lex situs* in the extraterrestrial realms. Other issues concern the untold mineral riches of other worlds, projects existing for the establishment of extraterrestrial mines. Are the extraterrestrial minerals "for taking" and, if so, would the first to come be also first in right? May land in outer space be occupied and subject to exclusive control? Other projects anticipate the permanent human settlement of the Moon and Mars, colonies evolving out of temporary facilities. Ideas have been advanced for the building of lunar hotels; blueprints for the utilization of vast lunar areas as solar energy collectors have been presented, as well as projects for the conversion of a lunar crater into a radioastronomy observatory. Most important, Dietrich and Goldstein (1998, p. 9) consider obvious that, absent an answer to the question of "whose writ runs up there", not many commercial activities will occur in outer space, being "irresponsible to encourage space development without some solution". They believe that the developed states will not advance the huge amounts necessary for developing the space resources while the developing countries claims on those benefits remain unsettled.

As seen from the outline *supra*, a thorough analysis is needed regarding the extent of permissibility and viability of property rights in the extraterrestrial realms, especially concerning the property status of landed extensions. This book will attempt to offer an answer to the above questions.

2.3 The Material Sources of Landed Property Law in Outer Space

Law – including the law of space property – is caused and shaped by several external realities; some of these are causative, representing the rationale behind the enactment of a certain norm (why law), while some are formative, representing the factors that give the law a particular shape and direction (how law). A clear distinction between these two categories of factors is not always possible; the technological, political, economic, cultural, social, financial environment often account for both the creation and shaping of legal rules and institutions.

Very often, the material sources of law are mentioned in the preamble of a legal document. An examination of this part of the 1967 Outer Space Treaty brings to light the main reason for its drafting, namely the great opportunities brought forth to mankind by its entry into the celestial realm. On a political level, the signatories agree that all mankind has a common interest in furthering the peaceful exploration and use of outer space, and that this activity needs to be carried out for the benefit of all peoples, regardless of their degree of economic or scientific development. The State Parties express their desire to partake in the international cooperation in this activity, in both its scientific and legal aspects, and state their conviction that such cooperation will strengthen the friendly relations between States and peoples, and will bring forth progress in their mutual understanding. Last but not least, the signatories explain that an Outer Space Treaty is bound to foster the principles and aims of the UN Charter.

Far from being empty rhetoric, the Preamble of the 1979 Moon Agreement embodies the rationale behind its drafting. The State Parties note the recent achievements of States in the exploration and use of the Moon and other celestial bodies, and recognize the important role to be played by Earth's natural satellite in the exploration of space and the benefits which may ensue from the exploitation of its natural resources. Bearing in mind the further development in the exploration and use of the extraterrestrial realms, the State Parties weigh up the existing *corpus juris spatialis*, and assert the need for defining and developing these international instruments. The signatories express as well their intent to ward off the Moon from becoming an area of international discord, being instead resolute to advance, on the footing of equality, the further progress of inter-State cooperation in its exploration and use, as well as that of the other celestial bodies.

Whereas this work is primarily aimed at restating the positive law of landed property in outer space by examining the forms under which legal norms manifest themselves, due consideration has to be taken of the factors that have created and shaped these norms. An outline of these agents follows.

2.3.1 Technical Progress as a Material Source

In a highly technological discipline as space law, science and technology play a leading role as agents of legal change. Technological changes were remarked by Goldie (1969, p. 153) as the occasion of contemporary evolutions in institutional and personal relationships, as well as law; the hydraulic engineering achievements allegedly determined the laws of land ownership, use and planning in ancient Egypt and China, whereas shipbuilding and navigation sciences contributed at the forming the law of the sea, as astronautic science did in forming the law of outer space.

The emergence and evolution of space technology is the main factor that summoned the birth of space law. If the Code of Hammurabi did not regulate stem cell research and the Laws of Manu did not govern the online transfer of currency, this is because no such activities occurred in ancient Babylonia and India. The meagreness of extraterrestrial property law can be thus explained by the novelty of space exploration. Decades ago, William Ogburn's "theory of cultural lag" postulated that the pace of technology exceeds the ability of the society to assimilate and manage it (The Texas Docket, 2000). This would mean that the legislator is unable to cope with the challenges of space, hence the extraterrestrial property *hypolexus* – the lack of sufficient legal norms.

Is there a cultural lag in the field of extraterrestrial space (property) law, or the law is actually ahead of facts? And should in fact law anticipate, or follow technological progress? As a general rule, the law has its origins from the facts (*ex facto sequitur lex*), yet a number of authors consider space law to be an exception. Stephen Gorove (1992, p. 78) noted that, while law is normally slow to react to societal changes, this has not yet been the case in the field of space law, although the pace is likely to diminish in the future. Sterns and Tennen (1992, p. 499) remarked the direct relation between the scientific and technological capabilities and the development of space law, noting that law in this field often preceded technology. In the year of the first manned Moon landing, Laszlo Szaloky (1969, p. 176) was able to "state with satisfaction that the principal rules relating to human activities on the Moon and other celestial bodies had been elaborated before man even set foot on them", and Oscar Fernandez-Brital (1969, p. 165) boasted that "space law never was behind technique, on the contrary, it was in advance with its concepts, to the achievements of technique". This phenomenon has been explained by Magno and Verdacchi (1976, p. 323) as being rooted in the perception of a rapid progress expected to arise from space activities, apparently having induced lawyers to rather prevent than follow out of fear of arriving too late, "when the matter of regulation had already gotten out of hand". Vassilevskaya (1980, p. 79) defends thus the Soviet proposal of 1971 for a Moon Treaty, notwithstanding the scepticism at that time that the document would be premature in setting legal regulations for activities feasible "perhaps in the distant future only". She explains that, while her country was aware of this, such a document was deemed as essential, serving the task of "precluding international conflicts as a result of lunar activities by various states and ensuring the legal basis for a possible utilization of the Moon" as stated by the Soviet Foreign Minister Gromyko. This paradigm change in the field of space law was advocated by Aldo Armando Cocca as early as 1958, especially regarding the legal nature of the Moon:

> In contrast to past methods of procedure, law today must anticipate the technical progress, and foresee the legal implications. The consequences which follow from a *fait accompli* cannot be permitted in the present stage of the development of civilization (Cocca, 1958, p. 34).

Forty years later, O'Donnell (1998a, p. 23) called, at his turn, for anticipating – if not entirely deciding – the plan for space development, "well in advance of settlers arriving in space". In practice, Article 11.5 of the Moon Agreement urges States Parties to establish a regime to govern the exploitation of lunar resources "as such exploitation is about to become feasible", whereas Article 18 calls for the implementation of the above norm "taking into account in particular any relevant technological developments".

Stowe (quoted by Moore, 1978, p. 277) and Von Rauchhaupt (1969, p. 201) however oppose law taking the initiative before a thorough analysis of the factual problems; Stowe exemplifies with the efforts to develop a definition for outer space as being – at that time – "a solution in search of a problem". Gyula Gal (1966, p. 55) also considers that suppressing the excess of zeal is an indispensable factor in the evolution of space law, calling the legislator to use restraint in considering unreal facts found in science-fiction novels and thus regulate "issues that will only be the task of future centuries or millennia".

Whilst in the past the regulation of space property rights may have been such a "solution in search of a problem", the legal process in the area of space law appears to have reverted to the classical paradigm. Three decades after the entry into force of the Outer Space Treaty, Francis Lyall (1998, p. 130) lamented that "[t]oo often the law has lagged behind the entrepreneur and the scientist", while Esquivel de Cocca (1997, p. 84) deplored that "law hurries up after technology, trying to adjust its provisions to facts already occurred", contrasting with the "previewing characteristic that Space Law used to have in its beginnings". She considers that some adjustments to the 1967 Treaty are necessary after three decades, given the "development of space activities, consequently the new situations, needs, factors and resources arisen". Szaloky (1969, p. 178) was foreseeing that, once the exploitation and use of the celestial bodies becomes more intensive and widespread, the rules of international law will have to be supplemented by civil law rules, while Vassilevskaya (1984, p. 275) believes that the building of large space structures on the Moon will lead to new types of activities, whose legal implications are not covered by the adopted agreements. She reckons this will call for reconsidering a number of questions which, although dealt with in the current law, will not be adequate in the future when large groups of people and settlers will be simultaneously present on the Moon.

Yet, the present *corpus juris spatialis* has not been drafted with the aim of comprehensively regulating private uses of outer space, though modifications and clarifications of existing space law will clearly be necessary as the commercial space age develops (Sterns et al., 1996, p. 60). According to Miklody (1979, p. 179), the Outer Space Treaty did not have as an aim the establishment of norms for the exploitation of the extraterrestrial minerals, given that the issue of comic mining seemed an "utterly utopian" fact at the time of the drafting of the treaty. Nonetheless – said she at the time – it is worthwhile to "test the problem *de lege lata*, and to prepare suitable principles *de lege ferenda* in the case that this utopian fact would become a reality and cause widespread opposition". Nauges (1979, p. 269) calls for urgently resolving a number of legal issues in view of the rapid progress of space technology, *inter alia* the problem of ownership in space. He believes that the Outer Space Treaty reflected the state of technology of that time, while its "imminent and rapid advance was either not foreseen or deliberately ignored", while the pace of technology accentuates the contradictory legal rules of space law. In his view, the evolution of space technology does not create new legal problems – it amplifies and unearths issues that were either overlooked, or shelved for political reasons, when the document was drafted.

Von der Dunk (1996, pp. 5–6) urges a balanced approach. Concerned that law may become a nuisance and an obstacle to progress, he would like to prevent it from running too far behind the day-to-day developments. Fearful too that law may become irrelevant since the future cannot be known, he would like to stop it from running too far ahead and from dictating "inflexible blueprints for the future". He prefers "the law as it stands today" as providing a "largely satisfactory regime for the time being", whilst the creation of a new regime from scratch is likely to give birth to more problems than it would solve. Myers (1975, pp. 71–72) too remarked the change of paradigm between *a posteriori* and *a priori* formulation of law in general, and space law. He warned that the development of specific space law norms prior to necessity is highly speculative, given that norms formed prematurely, at this stage of space exploration, may no longer be meaningful when applicable in practice; their details would "most certainly be subject to alteration as unknowns became knowns and as these related to the changing aspects of power politics on earth". He warned, in the same time, against developing specific regulations only after the need arises; agreement – says he – will be less successful due to the sharpening of the differing interests of the actors.

The question settling the dispute is whether the factual situation on the extraterrestrial realms calls for legal regulations in the field of property rights, and whether there is any fear of a *fait accompli* that would take the legislator unprepared.

2.3.2 Nature as a Material Source

The age-old conflict between the naturalist and positivist schools finds a special dimension in outer space as an area until recently untouched by positive law. Are legislators independent in creating the law of extraterrestrial landed property, or is there a superior order having already established unchangeable rules?

Natural law has been defined by Myers (1979, p. 185) as an absolutely just and immutable system, with universal and eternal validity, established by God or Nature; this law can be deduced from nature through the use of reason, and laws are the result of such attempts by man to reason from natural law. "It has been said" – writes Sir Paul Vinogradoff (1959, p. 169) – "that the law of nature is 'jurisprudence in the air'". He counters this statement by remarking that, "after all, the air constitutes one of the most important elements of life".

Can nature or God provide the answer to the question of "who owns the Moon"? Is natural law in outer space "jurisprudence in the void"? While Myers (1979, p. 185) questions the effective use of natural law as a basis for "intercosmic relations", others beg to differ. Cocca (1962, p. 2) regards the natural law "of the physical submission of the Moon to Earth" as establishing the submission of our natural satellite to the legal order of the planet. This basis – provided "by Nature itself" – can be "validly sustained against any eventual non-earthly legal order". A similar reasoning has been used in 1976 by the signatories of the ill-fated Bogota Declaration, who considered that, as a "physical fact linked to the reality of our planet, . . . the segments of geostationary synchronous orbit are part of the territory over which Equatorial states exercise their national sovereignty" (para. 1).

Divine law is recognized by Von Rauchhaupt (1969, pp. 202–206) as one of the forces that form and create the law of outer space; as divine law cannot be altered by human will, the space vehicles need to "adapt themselves to this space law that has been given to the celestial bodies". Regarding extraterrestrial landed property law, the Judeo-Christian divine law provides the very answer to the basic question of this book. God is the ultimate owner of everything: "The land must not be sold permanently, because the land is mine and you are but aliens and my tenants" (Lev.25:23, NIV). The New Testament is also inspirational for the stewardship of the extraterrestrial realms, the active aspects of the trust finding support in the parable of the talents, as it will be shown *infra*. Andem (1999) uses divine law as an aid to the objective interpretation of controversial concepts and terms of international law instruments. Thus, remarking the lack of a consensus in interpreting the Common Heritage of Mankind principle, he calls for using the Holy Bible as the "universal Code of Moral Conduct and Ethics" in solving the problems concerning the legal status of the celestial bodies. He thus finds justifications for the CHM concept in the Book of Genesis, where:

> [i]t can be seen . . . that we are all heirs and heiresses. . .[;] The sun, the moon, stars and other celestial bodies . . . were created by the Creator for the good of mankind as a whole (Andem, 1999, p. 10).

It will be nevertheless shown *infra* that the same natural law led John Locke to reason that exertion of labour upon these natural resources – originally bestowed by God in common upon all humankind – results in private property. It can be seen that the main weakness of natural law is that it generates different interpretations: what is natural for some may well be unnatural for others, and different systems of divine law offer conflicting answers. For instance, plans for building lunar surface cemeteries (Hofmann, 2000) may be challenged by the Navajo nation, given its protests at the placement of human ashes on the Moon, a *res sacra* to them (Associated Press, 1998). Religion proves thus to be a factor in the space use equation, including that of extraterrestrial landed property rights.

Nature shapes law also by providing environmental constraints in whose ambit law must evolve. Csabafi (1971, p. 28) opposes the "unqualified extension of international law into space environment", urging instead its interpretation and development "in accordance with the message the space medium conveys to us". Whereas Aristotle (quoted by Rubin, 1997, p. 7) spoke of a natural part of political justice with a universal force – "fire burns both in Greece and in Persia", truth is that fire does not burn the same on the Moon. The laws of nature have different incarnations in the extraterrestrial realms, presenting different challenges to *homo cosmicus*. In outer space, there is no "up" and "down"; on the Moon, there are no continents; the gas giants may not have a solid surface, hence the notion of "land" is warped;

water may be abundant on earth, but very scarce on the Moon. Legal principles and legal logic perfectly valid on earth may not be valid in the extraterrestrial realms. When the laws of nature are different, the laws of human interaction need to adapt to these new environments. It may well happen that the extraterrestrial realms will witness the development of new legal institutions, given the different environment and needs.

There are several environments similar, in some extents, with outer space. Reasoning by analogy has received different treatment from various authors, ranging almost from total rejection to total identity between the extraterrestrial realms and Antarctica, the High Seas, Spitzbergen and the airspace. Similarity alone does not necessarily impose a similar legal treatment, and the lawyer may chose to adopt completely different rules. The geographical-environmental situation of the UK and Japan may be extremely similar, though this does not mean that the law of one is, or shall be, applicable in the other; in fact, their legal systems have evolved in different directions. The extraterrestrial realms are a *sui generis* phenomenon. One may find inspiration in analogies, but these must be applied *cum grano salis*. Danilenko (1998, p. 252) narrates how the developing countries pushed for applying to the extraterrestrial natural resources, *mutatis mutandis*, the same Common Heritage of Mankind regime that has been established for the deep seabed. This approach has been opposed due to the technological and environmental differences between the deep seabed and the Moon and their respective resources, these differences precluding the automatic transplant to space law of norms and principles from other branches of international law (ibid., pp. 253–254).

In the same time, the conditions of the "space frontier" may influence the law – as the American frontier did not so many centuries ago. Historian Frederick Jackson Turner (1893) saw the American frontier as a material source of law:

> Behind institutions, behind constitutional forms and modifications, lie the vital forces that call these organs into life and shape them to meet changing conditions ... The legislation which most developed the powers of the national government, and played the largest part in its activity, was conditioned on the frontier.

Whether natural law may or may not provide the details for extraterrestrial ownership, it provides the ethos behind the *animus* of settling and owning the extraterrestrial realms. The law of nature has put into human mind the need for reproduction, hence human desire to colonize outer space. Aristotle's "naturalist" model of the legal order grounded in the animal nature of mankind, its inherent need to have children (Rubin, 1997, p. 7), has found its space-age order in mankind's need to leave its terrestrial cradle.

2.3.3 Actors as a Material Source

According to Helen Silving (1968, p. 257), the political and social climate of a particular country at a particular period determines the degree in which natural law is incorporated into positive law. Myers (1979, pp. 185–186) questions whether man will ever be able to fully understand and apply natural law; the abundance of legal and political differences among nations means that a common approach to law is far from realty.

Many authors saw the entry into Outer Space as giving the lawyers the unprecedented chance to build a "*Jus Novum*" (Magno and Verdacchi, 1976, p. 323), a "completely new and original" legal discipline (Magno, 1972, p. 165), to "give birth from nothing to a harmonious juridical corp . . . clearly applicable to poor and rich, to powerful and tiny States" (Hervy, 1972, p. 145). Maurice Andem (1999, p. 12) commends Jenks' stance against "imprison[ing] the development of space law in concepts and prejudices derived from an earlier stage in the development of international law". Shiffer and Snyders (1967, p. 239) urged legislators to "avoid burdening Space with the lengthy legal terminology of manifestly outdated earth law with its complicated systems and customs". Welf, Prince of Hanover (1961, pp. 126–127) also favoured keeping an open mind to new solutions in the juridical field and for entering the new realm of space "without prejudice or preconceptions, for this would only produce destruction before our conceptions have ripened from theory to knowledge". Andrew G. Haley (quoted by Hervy,1972, p. 148) also opposed the extraterrestrial prolongation of an anthropocentric law with its imperfections, limitations, conflicts and inconsistencies, as being the worst act that humankind could accomplish in its relationship with a universe that offers extraordinary prospects for a new era of peace and cooperation.

Myers (1975, p. 71) rebuffs the above logic; in his view, space law ought to be "anthropocentric" – in other words, it must be based on international law which is rooted in international politics:

> To believe that a new, separate or special law for outer space can be developed apart from national interest considerations is naive, unless one accepts the assumption that full-faith cooperation in space is in the national interests of each state and that national leaders can agree and act accordingly.

Indeed, the multitude of legal regimes is due to the variety of actors having shaped these rules, each pursuing their own interests. Positive law is not a phenomenon independent from society; it is a product of society and politics, and trying to separate these entities would be unwise. In the extraterrestrial realms, *corpus juris spatialis* is the result of antagonistic forces; the left/right, north/south, spacefaring/non-spacefaring, state/free enterprise divides are sources of law in their formative sense. Nauges (1979, p. 269) rightfully considers the rules of the Outer Space Treaty as "the fruit of compromise and hard-fought agreements".

One of the forces having stakes in the shape of space law – including extraterrestrial property rights – is private enterprise. At the beginning of the Space Age, there were calls for limiting the exploration of the extraterrestrial realms exclusively to States; this proposal was "called as unwarranted and unwanted and opposed" (Csabafi, 1962, p. 16). The concern of the entrepreneurial class with the legitimacy of its space activities was addressed by the Outer Space Treaty, who consecrated the freedom of private enterprise, as a nongovernmental entity, to conduct such activities under state authorisation and supervision. The space activities were initially limited in practice to those of States and their representatives; the situation slowly changed in view of the opportunities opening up before free enterprise (Gorove, 1983, pp. 200–204). In the first decade of the new millennium, the private activities in space took a giant leap, with the launch of the first space tourists, the building of the first private suborbital space vehicle capable of carrying humans, and of the first prototype for a private, inflatable space station.

The rather late emergence of the entrepreneurial class in the space equation explains both the current shape of the landed property law in outer space, and the need for a pro-active approach in this respect. According to Yasuaki Hashimoto (1993), the Outer Space Treaty and the related treaties pertain to a "first generation" of space law, treating the exploration and use of outer space by the public sector. During the cold war era, due to the huge expenses associated with the development and manufacture of space objects, most of the space activities were performed by public entities, and had either scientific or military purposes; profitability and commercial factors were ignored by national governments. The first generation of space law chiefly regulates the national activities in the exploration and exploitation of outer space, covering at the same time the private sector activities, but – says Hashimoto – this first generation may not adequately regulate all the future activities. Indeed, one has to agree – at the beginning of the space age, States were the main actors in the space arena, and law adequately regulated their conduct; private property rights were overlooked because there were not many private actors in space.

The larger the proportion of private activities in space, the larger the demand for a clear answer in the question of extraterrestrial ownership. In the same time, the bigger the involvement of private enterprise, the more is needed a pro-active approach towards supplying entrepreneurs with property rights. While Sterns et al. (1996, p. 60) consider the current law of outer space as neither expressly "anti", nor "pro" free enterprise, the US Presidential Commission concerned with the new space exploration vision expressed in 2004 differs. The treaty regime of the United States is seen as causing ambiguity in the legal status of a private company employing space resources in making products. Thus, it is found, the US have adhered to the Outer Space Treaty and, while not endorsing the Moon Agreement, neither did they challenge its basic assumptions and premises. Therefore, the Commission fears that this uncertainty could strangle in its cradle a nascent space-based industry, as companies will not invest important amounts of money in developing products to which their legal claim is unclear. The Commission agrees that, while the question of private property rights in space is highly complex, entailing legal issues at municipal and international level, it is crucial however to recognize and address these issues at the initial stages of the implementation of the vision. Otherwise, the Commission fears that the private sector activity associated with the space resources development – one of the key goals of the vision – will be insignificant (Aldridge, 2004, pp. 33–34).

Another force having a stake in the shape of space property law is the developing world. Whereas initially international space law was a product of the cold war and, basically, a bilateral phenomenon, the emergence of the third world countries transformed space exploration into an area of interest for all. According to Glenn Reynolds (1995, p. 119), many states with little stake in space exploration are involved in the consensus-based system of the UNCOPUOS, negotiations being plagued by "free rider" and "holdout" issues. Nonetheless, developing countries do have an interest in space exploration; their push for a New International Economic Order translated, in the area of space resources, into a demand for a share in these resources.

In drafting the law, O'Donnell (1998a, p. 29) outlines the need for balancing the conflicting interests of different factors – "national, international and space settlement interests, between exploration, development, and preservation goals". Whereas the interests of the private enterprise call for a conservative attitude, the interest of the have-nots call for a socialist attitude in the issue of space property rights. This antagonism between the "private property rights approach" and the "common heritage of the mankind approach" will be detailed later in this work. Reynolds (1995, p. 120) warns that, the more concrete the conflicting interests become, the more difficult it will be to negotiate law. While the present "veil of ignorance" makes agreement on general principles easier, the negotiations will likely become more difficult with the ability of the parties to clearly and specifically identify their future interests.

2.4 The Formal Sources of Landed Property Law in Outer Space

As with every human endeavour, space exploration does not take place in a social vacuum. Instead, the interaction between humans and the extraterrestrial realms is doubled by an interaction between humans regarding the same realms. By penetrating the celestial envelope, *homo cosmicus* did not cease being a *zoon politikon*; – it brought instead the outer space and the celestial bodies into the sphere of the law, giving birth to a new legal discipline – the law of outer space. Indeed, as remarked by Maria Picarel (1969, p. 189), law is "following the human as an inseparable and protective shadow wherever he goes, in space as well as on the Moon".

It is a recognized principle that the legal regime of immovables is determined by the law of the country where they are situated (North, 1979, p. 483). While every sovereign State is free to regulate the regime of the landed property located within its own territory, there are areas where no State sovereignty or territorial jurisdiction exists. As outer space and celestial bodies are not under the sovereignty of any State – national appropriation being outlawed by Article II of the Outer Space Treaty –, there is no municipal property regime to serve as *lex situs* in providing the answers to the questions of landed property rights.

Whereas no State has jurisdiction *ratione loci* (territorial jurisdiction) over the lunar surface proper, it has nonetheless jurisdiction *ratione personae* (personal jurisdiction) over its subjects, being entitled to regulate their conduct in the extraterrestrial realms, and in regard to the extraterrestrial realms. Thus, the 1967 and 1979 documents are not called "Treaty on the legal status of outer space" and "Agreement on the legal status of the Moon", but "Treaty on Principles **Governing the Activities** of States in the Exploration and Use of Outer Space, Including the Moon and Other Celestial Bodies" and "Agreement **Governing the Activities** of States on the Moon and Other Celestial Bodies". Even more than an entitlement, Article VI of the Outer Space Treaty compels its signatories to bear international responsibility for their national activities thereon, whether such endeavours are carried out by non-governmental entities or by governmental agencies. According to the same article, the activities of the non-governmental entities in the extraterrestrial realms are subject to authorization and continuing supervision by the appropriate State.

Property is a relationship between subjects in regard to a good, rather than a relationship between a subject and a good. Given the plurality of subjects and the absence of a unifying authority, the extraterrestrial realms can be seen as having a multi-dimensional legal status, depending on the subjects. Unlike municipal law, international law is consensual and is relative, treaty law applying only among its signatories. Among the States Parties to the Outer Space Treaty, the Moon has the status of international public domain. Among the States Parties to the Moon Agreement, the Moon has the status of the Common Heritage of Mankind. In fact, it is not the Moon that actually has a status as much as the different international actors chose to regard it in a way or another. In this way, it can be argued that *lex domicilii* – more precisely an international treaty that becomes part of the municipal law – has an important part to play in this particular situation.

As pointed out by Milton L. Smith (quoted in Wassenbergh, 1991, p. 82), the space powers may conclude an agreement between themselves, in which they recognize the validity of exclusive claims to the exploitation of the extraterrestrial natural resources filed by any other party to that agreement, such exclusivity having validity only among the parties. Indeed, given that eight Equatorial States have declared their sovereignty over some portions of the geostationary orbit in 1976 and that other States have agreed in 1979 to regard the Moon as the Common Heritage of Mankind, other States may chose to refrain to exploring and exploiting the Moon on environmental reasons, whereas other States may chose to mutually recognize their claims. All these agreements are, and would be, applicable only between their signatories.

Until then, Article III of the Outer Space Treaty directs States Parties to carry on their space activities "in accordance with international law, including the Charter of the United Nations . . ." Article 2 of the Moon Agreement, as well, provides for the carrying out of all activities on the Moon:

> in accordance with international law, in particular the Charter of the United Nations, and taking into account the Declaration on Principles of International Law concerning Friendly Relations and Co-operation Among States in accordance with the Charter of the United Nations, adopted by the General Assembly on 24 October 1970. . .

Article I of the Outer Space Treaty establishes the freedom of exploration and use of the extraterrestrial realms by all States "in accordance with international law", whereas Article 11.4 of the Moon Agreement ascertains the "right to exploration and use of the moon" by the States Parties "in accordance with international law". Article 6.1 of the same document guarantees as well the "freedom of scientific investigation on the moon by all States Parties ... in accordance with international law". Thus, it can be argued that international law becomes the quasi-municipal law for the extraterrestrial realms.

Public International Law was conceived to deal with relations between States; in this respect, unlike municipal law, it does not have the vocation to manage the property regime in the extraterrestrial realms. Still, even if not primarily aimed at regulating property, Public International Law has the capacity to do it; even more, some of its norms deal specifically with property law issues. The core itself of the Public International Law, for example – the *jus cogens* – prohibits slavery, that is, the "status or condition of a person over whom any or all of the powers attaching to the right of ownership are exercised" (as defined by the Slavery Convention of 1926). Legal norms also exist regulating matters of succession of States in respect of State property – i.e. the Vienna Convention of 7 April 1983 on Succession of States in Respect of State Property, Archives and Debts.

Having determined that the [quasi] *lex situs* in outer space and celestial bodies is the International Law, the next step is to examine the extent on which it succeeded in its regulatory mission. And we will begin by analysing the sources of the International Law, as listed by Article 38.1 of the Statute of the International Court of Justice (ICJ):

a. international conventions, whether general or particular, establishing rules expressly recognized by the contesting states;
b. international custom, as evidence of general practice accepted as law;
c. the general principles of law recognized by civilized nations;
d. subject to the provisions of Article 59, judicial decisions and the teachings of the most highly qualified publicists of the various nations, as subsidiary means for determination of rules of law.

These provisions are not to prejudice, according to Article 38.2 of the said Statute, "the power of the Court to decide a case *ex aequo et bono*, if the parties agree thereto".

2.4.1 International Conventions

Several international conventions exist in the field of space law, offering some guidelines if not formally establishing a property regime in the extraterrestrial realms. The 1967 "Treaty on Principles Governing the Activities of States in the Exploration and Use of Outer Space, including the Moon and Other Celestial Bodies", colloquially known as the "Outer Space Treaty" (OST) is very often perceived as a "Constitution" for outer space (Rosenfield, 1977, p. 440; Magno, 1972, p. 166), having set general principles to be further developed and not to be derogated from. Within a municipal law, a constitution is the supreme law of the country and the keystone to its legal system, having implications in all areas, including property rights. In this particular field, the OST outlaws national appropriation and sets several rules of use of the extraterrestrial realms, which will be examined throughout this book.

Although relevant, the provisions dealing with space property rights are in short supply in the OST. This is because the treaty, as a fundamental text, establishes only the main regulating principles that call for further elaboration once the exploitation of the lunar resources has begun, and does not legislate specially (Ferrer, 1969, p. 144; Szaloky, 1973, p. 197; He, 1997, p. 52). A number of conventions further its provisions, a particular position in the field of property rights being hold by the "Agreement Governing the Activities of States on the Moon and Other Celestial Bodies". The document, colloquially known as the "Moon Agreement" (MA), has been adopted by the UN General Assembly on 5 December 1979, and entered into force on 11 July 1984, following the deposit of the fifth instrument of ratification (MA, Article 19.3). The MA is the most comprehensive document outlining a [non-] property regime in Outer Space, K. Narayana Rao (1981, p. 278) considering its Article 11 "a complete code in itself on the question of exploitation of the Moon's resources". This is an overstated opinion, given that the article in cause calls for its further elaboration (Article 11.5).

The views on the MA vary, depending on the political persuasion of the evaluator, from its branding as "the archenemy of space development" (O'Donnell, 1998b, p. 1) to "the man who wasn't there" of Space Law. According to Glenn Harlan Reynolds (1995, p. 118), the MA is "pretty generally considered a failure, even by its supporters". By 2006, only 13 States have acceded to it (Australia, Austria, Belgium, Chile, Kazakhstan, Lebanon, Mexico, Morocco, Netherlands, Pakistan, Peru, Philippines, and Uruguay) – none of them a major spacefaring State. Three of the States Parties (Austria, Belgium and the Netherlands) are members of the European Space Agency, yet this does not impose any obligations upon ESA. This would happen only if ESA would declare its acceptance of the rights and obligations provided for in the MA, and if a majority of the ESA Member States would be States Parties to the MA and to the OST (Article 16, MA).

The ICJ statute requires the rules established by means of convention to be "expressly recognized by the contesting states", and cannot be applied to other actors – "*pacta tertii alieni nec nocere, nec prodissi potest*". Whereas the MA is a "victim" of the voluntary character of international law and cannot serve as a standard for the international community due to the small number of States Parties, its provisions are nonetheless binding upon these States under the principle "*pacta sunt servanda*".

As stated elsewhere, treaty law is not very clear in the field of space property rights. Daniel Goedhuis (1981, p. 1) considers that the current treaties in the field of space law contain ambiguous clauses, through which the drafters have put on hold the problems of resolving the disputes expected to arise in the practical application of space activities. Smirnoff (1972, p. 175) deems necessary the clarification of certain principles of the OST. Magno (1972, p. 166) recognizes that, whereas the OST is the constitutional text of outer space, it is not immutable, and it can therefore be modified and integrated. Indeed, Article XV of the OST authorizes any State Party to propose amendments to it, these entering into force for each accepting State Party upon their acceptance by a majority of the States Parties to the Treaty; an identical provision is contained in Article 17 of the MA. What matters is the contents of these amendments, strengthening or weakening the current regime of property rights in outer space. The Moon Agreement, in its Article 18, contains procedures for its possible review, "in the light of past applications of the Agreement". Although the norm called for including the question of the MA review a decade after its entry into force on the UN General Assembly agenda, no such action was deemed necessary. Calls have been made – as it will be seen *infra* – for an altogether radical measure, that of withdrawing from the OST. Article XVI of the document authorizes any State Party to withdraw from the Treaty by giving a one year notice. Article 20 of the Moon Agreement contains a similar provision.

2.4.2 International Custom

Numerous authors consider the basic tenets of the OST as forming part of customary international law (Matte, 1984, p. 318), having been established long before the conclusion of the treaty (He, 1997, pp. 53–54). As such, these bind all nations, and not merely the State Parties (White, 1991, p. 2; Wiewiorowska, 1979, p. 24). Lyall (1998, p. 131) considers that the various UN "Space Resolutions followed by compliant practice" may be "formative of customary International Law, albeit that the usual chronology of the creation process is inverted". The OST codified and further developed these customary law principles who were, according to Paxson (1993, p. 488), the first sources of space law. These include the freedom of use and exploration of space by all States, the non appropriation principle, and the maintenance of jurisdiction by states over objects launched into outer space, while Silvia M. Williams (1997, p. 178) considers that international cooperation is moving towards achieving this status.

According to Paxson (1993, p. 488), "no obligations to share space benefits emerged among the important customary principles that developed". Whereas the Outer Space Treaty codifies "international custom, as evidence of general practice accepted as law" (Article 38.1.b ICJ Statute), the claim that the Moon Agreement represents customary law is seen by Kurt Baca (1993, p. 1068) as "probably not credible". It will be shown later that several norms of the MA in the field of extraterrestrial samples, do actually represent customary law.

The status of the non-appropriation principle as a norm of customary law presents a special interest. Qizhi He (1997, pp. 53–54) points out the debates surrounding the legal status of the geostationary orbit, where a number of equatorial countries who were not parties to the OST considered themselves not bound by its principles. This argument, according to He, was rejected by a great majority of States, on the ground that the OST enumerates fundamental principles and rules representing the "existing general customary law which shall bind all members of the international community independent of a formally ratifying or accepting the Treaty". Baca (1993, p. 1068) quotes authors supporting the view that the OST "has become a statement of customary law" on the basis of the presumed "acquiescence of states to the principles embodied in the treaty before and after its ratification", so that while withdrawal will release states from norms of *jus dispositivi*, the norms of *jus cogens* will still bind these states. Francis Lyall (1998, pp. 130–131) hopes that a state withdrawing from the OST "would continue to consider itself, and be held by appropriate process, to be bound as a matter of customary law by the rule otherwise articulated in Article II". We see as debatable whether the non-appropriation principle is a matter of *jus cogens*, hence immutable.

Whereas treaty law can precede technological progress, customary law is dependent on actual practice. Edwin W. Paxson (1993, p. 490) considers the former, rather that customary law, as better suited for governing space activities for several reasons, one being that "the regulation of space activities develops more quickly than the actual practice of States in the exploration and exploitation of space". Wiles (1998, p. 516) anticipates, logically, that "customary law will grow and change as space access and use develops". Indeed, in 1998, US entrepreneur James Benson stated his intention to "fly a privately sponsored deep space science mission" and to claim ownership of an asteroid with the aim of creating "an important and historic precedent" (Benson, 1998) in support for space property rights.

2.4.3 The General Principles of Law

The "general principles of law recognized by civilized nations" are a nebulous legal category, their aim being, according to Hersch Lauterpacht, of authorizing the International Court of Justice "to apply the general principles of municipal jurisprudence, in particular of private law, in so far as they are applicable to relations of states". In the *Abu Dhabi* award given by Lord Asquith in 1951, reference was made to the "principles rooted in the good sense and common practice of the generality of civilized nations – a sort of 'modern law of nature'" (Lauterpacht, 1955, pp. 29–30). It is hereby offered that such a common sense principle relevant to extraterrestrial property law is the "first come, first served" doctrine.

Whereas the extraterrestrial realms have never been under the territorial jurisdiction of ancient Rome, many modern systems owe an important debt to Roman law – a system well suited to regulate complex legal transactions. Therefore, through their recognition by "civilized nations", many principles of *Jus Romanum* are a logical choice when considering the law of property in outer space. Private Roman law is anything but silent; therefore, reference to it is often made in the present work. While Roman law has been used by many scholars in analysing problems of space law, not everybody assents. Polish lawyers Hara and Stanczyk (1984, pp. 54–55) oppose the "mechanical transformation of concepts derived from the Roman law or principles accepted in other branches of international law". Use of Roman law concepts by the space law publicists is seen by the two authors as causing terminological chaos and misunderstandings. Another reason for countering the use of Roman law is the alleged recession it causes in the "progressive development" of space law, whereas they would rather see this field of law as drawing "from the most progressive tendencies in legal thinking and not from the narrowly interpreted concepts of state's sovereignty or national interests".

The "general principles of law" should not be confused with the United Nations-sponsored "principles of international law concerning friendly relations and co-operation among States" or "general principles of international law". Such concepts of general international law are, for instance, the "principle of peaceful settlement of disputes and the principle of non – aggression" (Matte, 1984, p. 285). Whereas several authors speak of the common heritage of the mankind as a "principle" of international law, Dekanozov (1977, p. 198) finds such an approach as being "ungrounded". Jose Montserrat Filho (1996, p. 60) believes as well that "we are still very far from having the 'view point of mankind' as a basic principle of international law". The "principle of co-operation and mutual assistance" is the only such principle expressly mentioned in Article IX of the Outer Space Treaty and Article 4.2 of the Moon Agreement as a guide for States Parties in the exploitation and use of the extraterrestrial realms, whereas Article 2 of the Moon Agreement provides for the carrying out of all activities on the Moon "taking into account the Declaration on Principles of International Law concerning Friendly Relations and Co-operation Among States".

2.4.4 Judicial Decisions

The ICJ Statute lists judicial decisions as a subsidiary means for the determination of the rules of law (Article 38.1.d). Until now, there are no judicial decisions either of the ICJ or of arbitrary nature that would specifically settle an issue of space property rights. Nevertheless, a number of judicial decisions delivered in other fields are relevant at the level of principle or as analogies – to be taken *cum grano salis*. In the same time, while the jurisdiction of the ICJ entails contentious issues between States, private citizens have brought the subject of space property rights to the attention of domestic courts. Such is the case of Adam Ismail, Mustafa Khalil and Abdullah al-Umari of Yemen, who in July 1997 filed a lawsuit in Sana'a, demanding the immediate suspension of all NASA operations on Mars on the grounds the Red Planet is owned by them (CNN, 1997). The plaintiffs withdrew their case after being threatened with arrest by Yemeni Prosecutor General Mohammad al-Bady, who declared they were only seeking publicity and fame (Reuters, 1997).

Another frivolous suit was brought against NASA by a Russian astrologer. Learning that the "Deep Impact" mission would create a crater on Comet 9P/Tempel 1, spiritualist Marina Bayross came to see the scientific experiment as an encroachment upon her system of values and as a disturbance of the natural balance of forces in the Universe (Bayross, 2005). Consequently, in March 2005, the astrologer filed a lawsuit with the Presnensky District Court in Moscow against NASA, demanding that the space agency call off the mission. Whereas she did not formally claim that the comet was her property, by seeking to control NASA's actions towards that body she did advance such a claim, under the thesis that property means control over access. The plaintiff also sought the equivalent of the entire cost of the mission in compensation for moral damages. While the Presnensky District Court dismissed the initial case on account of NASA's immunity from Russian jurisdiction, the Moscow City Court took up the appeal (Arutunyan, 2005) and, following a hearing on May the 6th, it cancelled the previous dismissal as Bayross' lawyer pointed out that NASA's office in the premises of the US embassy in Moscow does fall under Russian jurisdiction. The higher instance ruled that Bayross could proceed with the lawsuit (Agence France-Presse, 2005), sending back the case for further consideration with NASA's participation. Upon several postponements due to the absence of representation from NASA, the claim was eventually dismissed on November 8, 2005 (RIA Novosti, 2005).

The most relevant court case in the matter of space property rights was brought by Gregory W. Nemitz, a space activist and owner of Orbital Development, against the United States. The judicial stage of the affair would follow a lengthy exchange of letters between him and several institutions, as detailed *infra*. On March 3rd, 2000, Nemitz had claimed ownership of asteroid Eros with the Archimedes Institute, considering his claim for "a single asteroid the size of a mountain" – smaller than many private properties on Earth – as "reasonable" when compared with larger extraterrestrial claims. Considering space property rights a chief issue in need of resolution, Nemitz was looking for ways to address this matter. When in February 2001 the NEAR Shoemaker spacecraft landed on Eros, Nemitz sent NASA a $20 invoice for parking their spacecraft on his property, setting a low amount as a proof his goal was in fact to bring publicity to the issue of space property rights (Gregory W. Nemitz, personal communication, November 28, 2005).

Despite the symbolic amount requested, NASA's General Counsel Edward A. Frankle replied with a letter dated 9 March 2001, in which it declined to make the requested payment. Frankle explained that Nemitz failed to cite:

> any legal basis for Orbital Development's claim to own Eros or to any legal significance of a filing with the Archimedes Institute. ... If Orbital Development or its principals are U.S. nationals, [the Outer Space Treaty] would seem to preclude any claim to own Eros.

Frankle volunteered to forward any material that OrbDev could provide in support of their claim to the Department of State for its advice and guidance.

Nemitz did oblige, supplying NASA a few weeks later with the arguments he deemed appropriate for defending his extraterrestrial ownership. "Multiple facets of the law allow support of the claim", wrote Nemitz, agreeing though that there is "difference of opinions on the Space Property Rights issue". According to Nemitz:

> [a] claim on an un-owned thing immediately converts the thing into property of one species or another. The claimant is the original owner of that property.

Moreover, considered he, the claim can be sustained with him being "the 'first to sight the opportunity' of establishing a valid and viable mining claim", NASA being prohibited by law from doing so while "mining claims by individuals are not prohibited". Another argument put forward by Nemitz is that:

> Physical possession is not the sole ... basis for property ownership. The pursuit of a thing with intention of possession is the basis for one species of property right. A thing being used, and/or with the infusion of one's own efforts into the thing, creates another specie of property right. The existence of either of these activities does by deed convert the thing into real property in the eyes of the law.

In the same time, he argued that his claim has "a higher standing than any other that can be possibly brought forth" pending "some entity tak[ing] physical possession of 433 Eros and attempt[ing] to eject [OrbDev's] claim by their usurpation or squatting". Therefore, for the moment, he argued, "the claim exists", his "full property rights remain valid", and his claim is "absolutely not premature or inappropriate" (Letter from Gregory W. Nemitz to Edward A. Frankle, 22 March 2001).

On April 9, 2001, NASA's General Counsel Edward A. Frankle replied to Mr. Nemitz, informing him once again of his failure to cite any foundation in law for the claim to own Eros, and denying again his "request for payment of a 'parking/storage fee'". In considering so, NASA saw no need to take any position on the relevance of the 1967 Outer Space Treaty to private individuals, considering instead the claim as depending on the establishment and validity of OrbDev's ownership of the asteroid in cause. On January 21st, 2003, NASA's E. Jason Steptoe sent their final determination to OrbDev, stating that as nothing has changed, there is no need for consultation with other federal agencies, and the matter was now closed.

On February 13, 2003, Nemitz, in his quality of "aggrieved citizen", sent a notice to Secretary of State Colin Powell stating that NASA has exceeded its authority, to which the Department of State replied on 15 August 2003 through a letter from Ralph L. Braibanti, Director, Space and Advanced Technology, US Department of State/Bureau of Oceans and International Environmental and Scientific Affairs:

> In the view of the Department, private ownership of an asteroid is precluded by Article II of the [1967 Outer Space] Treaty ... Accordingly, we have concluded that your claim is without legal basis.

Nemitz deems this particular official action to have created a "'case in actual controversy' which gives a Federal Court jurisdiction to hear the case". Consequently, in a press release dated August 25th, 2003, OrbDev stated that it "continues to dispute this controversial conclusion by the Department of State and will file suit in Federal court for a definitive decision that will be binding on the U.S. government". On November 6th, 2003 Nemitz filed in Federal Court in Reno, Nevada the matter concerning "*Treaty vs. the Natural, Inherent Rights of Man*" to acquire and own property. On January 28th, 2004, the US Attorney's office filed Defendant's Motion to Dismiss, arguing that the Plaintiff did not state sufficient material facts to support his arguments.

On February 11, 2004, Nemitz conceded in his Response to Defendant's Motion to Dismiss that his claim for "breach of implied contract" was insufficient to be sustained. While that meant that the US Government did not have to pay the fees requested by OrbDev in relation to the spacecraft storage, the plaintiff maintained that his request for a Declaratory Judgement concerning the rights of individuals to own private property in outer space remains viable, having to be decided by the court. Nemitz requested an oral hearing on the matter of the Motion to Dismiss.

On April 26th, 2004 U.S. District Court Judge Howard McKibben ordered the case to be dismissed. "A complaint may be dismissed as a matter of law for lack of a cognizable legal theory", read the Order to Dismiss, further stating:

> A takings claim under the Fifth Amendment of the US Constitution requires a constitutionally protected property interest. ... Nemitz has failed to assert such an interest. Neither the Ninth nor the Tenth Amendments provides a cognizable cause of action for the denial of a property interest in Outer Space.

Unhappy with the decision, Nemitz filed an Appeal Brief in the United States Court of Appeals for the Ninth Circuit. The Court issued, on February 10, 2005, a Memorandum in appellate case No. 04-16223, *Gregory William Nemitz vs. NASA*, et al., affirming the lower court's dismissal for "failure to state a claim upon which relief can be granted." When the 9th Circuit Appellate Court reaffirmed the lower court's decision, Nemitz renounced taking the case to the Supreme Court, and considers the case closed (Gregory W. Nemitz, personal communication, November 28, 2005). Whereas the legal arguments for dismissing Mr. Nemitz's claim have been outlined by us *supra* ("claiming does not mean owning"), it is notable that, by taking the matter to an official level, Gregory Nemitz opened a door where one can glimpse the shadow of the trials to come, when humankind will return to the Moon and go further, colonizing the solar system.

2.4.5 The Teachings of Publicists

Although only a subsidiary means for determination of legal rules, the teachings of publicists are, quantity-wise, the richest source of law in the field of space property rights. The scholars involved in the study of space law are indeed some of the "most highly qualified publicists of the various nations" as required by Article 138.1.d of the ICJ Statute. The "Proceedings of the Colloquium on the Law of Outer Space", "Journal of Space Law", "Space Policy" and other publications contain some of the finest analysis on the area of extraterrestrial landed property rights. In addition, several private and semi-private legal groups have passed resolutions addressing, *inter alia*, these issues. Reference to these and many of the works of publicists is made throughout this book.

Regardless of their fine quality, the teachings of publicists remain nevertheless a secondary legal source. The space property rights field has attracted scholars of widely differing intellectual perspectives. The law has to be clear and unequivocal; the opinions of one publicist are different from the opinions of another one, and many times they express completely opposite views. If one author justifies the private appropriation in outer space and another one opposes it, which view is legal and which is not? The publicists only interpret the law and attempt to clarify it – they do not create the law, at least not directly. However, many of their *lex ferenda* proposals end up by being adopted as *lex lata*. Edward R. Finch (quoted in Moore, 1978,

p. 279) has remarked that it is very often that papers written in the field of space law become the actual position of States in the UNCOPUOS, cautioning American space lawyers:

> We must be careful in what is academically proposed toward legal interpretation of outer space regimes because other states have started to do their academic legal interpretation and planning on an organized basis.

In 1921, Lord Sumner (quoted by Csabafi, 1971, p. 26) declared, in the case of "*The Kronprinsessan Margareta*" (1921) that "the more the field is covered by decided cases the less becomes the authority of commentators and jurists". Whereas judicial decisions in the field of extraterrestrial property rights do not abound, one is entitled to paraphrase Lord Sumner in stating that, the less the field of space property rights is covered by decided cases, the more becomes the authority of commentators and jurists.

2.4.6 The Completeness of International Law

While in 1978 NASA's General Counsel Neil Hosenball (quoted in Moore, 1978, pp. 276–277) was hoping that *hyperlexus* – a "pathological condition caused by an overacting lawmaking organ" – would not become virulent among space lawyers, nowadays, international space law seems to be on the contrary the victim of *hypolexus* – the lack of sufficient legal norms. The depths of space are doubled, in Kenneth Silber's (1998) view, by an "even more forbidding" realm: the "uncharted legal territory, and unpredictable politics, of owning property out there". He remarks the "general silence of national and international law on extraterrestrial property rights", this field being in his view "murky at best, downright hostile at worst". Roberts et al. (1996) see the extraterrestrial realms as harbouring "a legal vacuum almost as complete as the vacuum of space itself". They deem the present space law as "rather uneven", its principles "often retreat[ing] into ambiguous generalities or leave gaping holes which must be filled before any serious commercial development can occur" As for domestic standards, they are considered "either absent or so cumbersome as to inhibit exploitation". H. Nauges (1979, p. 269), at his turn, voices a very genuine concern:

> The difficulty is not . . . that we are starting from a legal vacuum but that there are a number of abstract, imprecise, insufficient and sometimes contradictory legal rules which are likely to be subject to genuinely differing legal interpretations.

Magno (1972, p. 166), on the contrary, saw Space Law to be the clearest of all the legal branches and of all the legal disciplines. We respectfully disagree with the latter view. Treaty law norms on this field are, indeed, very scarce and when they do exist, they are imprecise. They fail to define concepts that are fundamental to the law of landed property in outer space, such as "celestial body". This particular definitional issue raises the question of ownership of asteroids: are they celestial bodies and, hence, outside the sphere of appropriation?

Silent or obscure as it may be, the absence of a "lunar civil code", or of clear rules of international law concerning landed property rights in outer space, should be of no consequence in practice. According to Hersch Lauterpacht, one of the most established rules of international law, firmly based on universal arbitral and judicial practice, is the prohibition of *non liquet* (i.e. "the law is not clear") or, in its positive formulation, the principle of the completeness of international law. This means that a court cannot decline to give a decision on the ground that an applicable rule of law is absent, or insufficient, or controversial, or uncertain and lacking in clarity (Lauterpacht, 1975, pp. 214–217). "Silent law", therefore, does not mean "no law", and legal lacunae need to be bridged. In fact, it is mostly treaty law that is silent or unclear; as seen *supra*, other sources of international law with relevance in regulating extraterrestrial landed property rights do exist. We may not pronounce, at our turn, a *non liquet*, and we are thus faced with the cabalistic task of reading between the lines of public international law norms, their implications on the private sphere, as well as analysing all the possible sources of international law.

In the same time, the absence of limiting rules of law may result in invoking what Lauterpacht (1975, p. 220) calls "the residuary rule of presumptive freedom of action", though he reckons that this rule is seldom, if ever, asserted. Goldie (1976, p. 289) does use this rule; in his view, no prohibitory norms exist in treaty or customary law that would forbid states or individuals to appropriate resources in "masterless areas", this action being also justified by the maxim "what is not prohibited is permitted". Von der Dunk (1999) as well would start any space law discussion with the fundamental rule concerning freedom of space activities, namely "everything that is not, one way or another, prohibited or conditioned, is allowed." There are however opponents of this principle; according to Markoff (1974, p. 136), contemporary international law does not consider the lack of prohibition as an authorization or permission, and claims that a given activity should be recognized as a permissible one because not expressly prohibited would get back international law with many decades, to the "voluntaristic sentence of the Permanent Court of the Hague on the The Case of the S.S. "Lotus" case" (1927). We respectfully agree with the classical school of "what is not prohibited is permitted", while being aware that prohibitions need not be always explicit (on the issue of imaginary "loopholes" in the OST that would justify private appropriation of celestial bodies, see *infra*).

2.5 Conclusion

As seen *supra*, the norms and rules regarding property rights in outer space and celestial bodies are limited in extent and in quality. There is a consensus among the space law scholars, practitioners and industry representatives that the way international law addresses property rights in space is ambiguous – as we will see further on, law fails to define essential concepts as "celestial body". Besides this, an increasing number of voices call not only for a clarification of the regime, but also for a positive approach towards property rights. These are seen as essential, being a source of incentive for space development both by securing investments and by diminishing conflicts among users. Whereas many issues in this field have been swept under the rug by past legislators, future treaty law needs to be clear. Otherwise, the insecurity will either preclude private exploration of space, or will put the international community in front of a *fait accompli*. The new trends in privatisation of space endeavours and the planned return to the Moon by 2020 shift the subject of property rights in outer space from the field of Byzantine discussions into the sphere of the concrete.

Chapter 3
The Object of Landed Property Rights in Outer Space[1]

Give me a lever long enough and a fulcrum on which to place it, and I shall move the world

Archimedes

3.1 Introduction

The Outer Space Treaty contains in Article II a fundamental principle, outlawing the national appropriation by any means of outer space and celestial bodies. Strict as it may be in this prohibition, the Treaty fails however to define the precise object of its application. This silence has prompted two disputes in the specialised academic circles: the legal definition of outer space, and the legal definition of a celestial body.

This chapter discusses whether [some] asteroids and comets are "celestial bodies", immovable land-like territorial extensions that cannot be appropriated under present regulations – or floating movable things, orebodies *ferrae naturae* capable of being captured and reduced into private ownership. Several theories are examined, such as the employment of the spatialist and functionalist approaches, or the use of the criterion of actual movability from orbit by human action. The present chapter weighs the different advantages and disadvantages of each of the above schools of thought regarding the definition of a celestial body, and attempts new approaches based on original findings, such as the analogy between the legal status of asteroids and icebergs.

One of the reasons precluding a proper legal definition for celestial bodies is the absence of a consensus in the scientific circles of what is – and what is not – a planet. The discovery of Kuiper Belt objects – one of them, Eris (2003 UB_{313}) being, in fact, larger than Pluto, has prompted a vivid debate among astronomers as to how many planets there are in our solar system. Whereas Pluto has been demoted to a "dwarf planet" in September 2006, not all astronomers agree or are willing to accept this; without a generally accepted scientific definition serving as a factual basis, it is likely the legal debate will continue.

[1] A version of this chapter has been published as Pop, V., 2001. A celestial body is a celestial body is a celestial body. *The 44th Colloquium on the Law of Outer Space.*

V. Pop, *Who Owns the Moon?*, Space Regulations Library Series 4,
DOI 10.1007/978-1-4020-9135-3_3, © Springer Science+Business Media B.V. 2009

While the question "how tall is the sky" has been brought to practice on the occasion of the Bogotá Declaration, the inquiry into the legal concept of a celestial body has been, until now, unsubstantiated by any actual need. However, the intention of the US entrepreneur Jim Benson (1998 pp. 46–49) to appropriate an asteroid by means of effective possession makes it an emerging issue of interpretation and application of the *Corpus Juris Spatialis.*

In advocating private property rights in outer space, some authors claim that the Outer Space Treaty outlaws only the national appropriation of celestial bodies. It will be shown in a following chapter that this is not the case, private appropriation being in fact denied the safeguarding shield of state protection. However, a different approach could prove more successful, that would make reference not to the contents of property rights, but to their object. According to this approach, private appropriation of [some] asteroids and comets would be allowed not because celestial bodies can be privately appropriated, but because [some] asteroids and comets escape the non-appropriation principle, being in fact not celestial bodies in the legal sense.

Claims of ownership over asteroids have already been made, such as OrbDev's claim over Eros followed by an invoice to NASA for a parking/storage fee. The triviality of extraterrestrial property claims unsubstantiated by any *corpus* is analysed *supra.* However, while challenging the private applicability of the non-appropriation principle, these claims do not dispute that they are directed at celestial bodies.

We would join other scholars in considering that some extraterrestrial resources are not, legally speaking, celestial bodies.

3.2 The Concept of "Res" in International Space Law

The concept of "things" is a very vague and heterogeneous notion. While general law employs terms as "goods" and "real estate" as species of "things", *corpus juris spatialis* uses special categories that have very ambiguous legal definitions ("space objects") or no legal definitions at all ("celestial bodies").

What is legally a "thing" in outer space? The universe is populated by astronomical objects that present an extreme variety, from black holes to quasars to nebulae to planets, moons, asteroids, comets, etc. Are all of these objects of rights – in other words, are they, things?

In defining what is legally a "thing", we would apply the theory of reasonable man. Recently, astronomers have discovered quasars more than 10 billion light years away, the farthest objects ever seen in space. These may be objects in the astronomical sense; we would however submit that they are not objects in the legal sense. Neither should nebulae, black holes, stars other than our Sun, or extrasolar planets, be considered as falling under the sway of *corpus juris spatialis*. While the Outer Space Treaty does not impose itself any territorial limits, it would be unreasonable to extend terrestrial law to the scale of the universe. The Moon Treaty, despite its poor record of ratification, does contain a reasonable limit, its provisions applying

"... to other celestial bodies within the solar system, other than the Earth ..." (Article 1.1.). It is thus submitted by us that the legal notion of "thing" does not have any validity beyond our solar system. Everything in the solar system is legally a thing: the planets, asteroids, comets, meteors, orbits, etc; beyond the limit of our solar system there is, legally, nothing. This is, naturally, valid at the present time. Humankind has not yet touched the stars; its most distant messengers, the Voyager and Pioneer spacecrafts, will be soon leaving the solar system without losing however their legal quality of things. However, for now, one should not try to regulate something one cannot yet reach.

This is already a generous limit – the solar system does not finish with the last planet but continues with the billions of comets orbiting the Sun in the Kuiper Belt which begins just beyond the orbit of Neptune, and the 2,000 billion of other comets located in the Oort Cloud that extends as much as a one and a half light-years from the Sun (Genta And Rycroft 2003, p. 268). In fact, every bit of dust that is in the gravitational hold of the Sun counts as part of the Solar System, so the outermost of such dust may reach half way to the nearest star (Ottewell, 1989).

Having established that everything within the solar system, but not further than that, is legally a thing, the next step is to analyse the legal sub-categories. What is a celestial body, and what is not, in our solar system?

3.3 Territorial Resources vs. Material Resources, Immovables vs. Movables

The object of property rights is constituted by "things" – *res* – which is a very vague and heterogeneous notion. The legal treatment applicable to various classes of things is fundamentally different, material extensions having a separate legal dimension from territorial extensions, and movables from immovables. As agreed by Burn (1994, p. 5), "[l]and and goods are and must ever be on a different plane".

In considering a question of a proprietary or a possessory nature, the first task of a court is to decide whether the *res litigiosa* is a movable or an immovable. Depending on this distinction is the legal system that will be applied to the case. It is a generally accepted principle of Private International Law that, while the *lex situs* -the law of the country where the thing is situated- regulates the legal regime of the rights over immovables (North, 1979, p. 483), its importance is diminished regarding the rights over movables; in this second instance, the *lex domicilii* has an important role to play according to the principle *mobilia sequuntur personam*(North, 1979, p. 552).

The only law that can determine in an effective manner whether a thing is to be treated as a movable or as an immovable is the law of the state that has control over the thing, i.e. the *lex situs* – the law of the country where the thing is situated (Collins, 1987, p. 899). As shown in the previous chapter, the non appropriation principle implies that there is no municipal property regime to serve as *lex situs*, this role being fulfilled by international law, more precisely by space law. However, *corpus juris spatialis* does not refer precisely to movables and immovables, making instead use of cognate legal categories, namely *celestial bodies*, *space objects*, *resources*, *samples*, *orbits*, etc.

Determining what constitutes in law a celestial body is therefore central for the study of property rights in outer space. Pursuant to the non-appropriation principle of Article II of the OST, celestial bodies cannot be appropriated. In practice, should [some] asteroids and comets be considered celestial bodies, they would fall under this prohibition; *per a contrario*, if they are not celestial bodies, they may become the object of private property rights.

While the terms "land" and "goods" are not used in space law, there exist nevertheless legal categories that can be categorized as spatial extensions and others that are material extensions; therefore, there are legal categories regulated by *lex situs* and legal categories regulated by *lex domicilii.*

Land is the archetypal spatial extension, as opposed to a material extension. Should one take away the substance of the land, the spatial value still remains. One cannot consume land; it may, at worst, make it unsuitable for use, but it cannot completely destroy it. Should one dig a hundred kilometre hole in the ground and take away all the mass, there will still be the space of the land that is left. The same, an orbit may become unsuitable for use by accumulation of debris, but it does not physically disappear. While outside the legal profession landed ownership seems a flat concept, in fact landowners do not own surfaces; they own pyramidal entities that may be x metres long, y metres wide, and about 6,378 km deep – the radius of Earth from the surface to its centre – *usque ad inferos* –and some kilometres high, represented by the height of the atmosphere – *usque ad coeli.*

Celestial bodies proper, orbits, points in space and outer space proper are spatial extensions. Outer space and orbits are purely spatial extensions, as they do not have any material existence. Unlike incorporeal things of relevance in space law such as intellectual property rights and the frequency spectrum, they do exist in three dimensions and, in the absence of the non-appropriation principle, they could be brought under the sway of national territorial jurisdiction. Such was the case with the geostationary orbit, which was declared in the 1976 Bogota Declaration to be part of the national territory of several equatorial States (see *infra*). Celestial bodies, outer space and its sub-categories (orbits and point positions) have characteristics analogous to the municipal category of immovables.

Other extensions in outer space have characteristics analogous to the municipal category of movables. Such is the case, for instance, with space objects that are regulated by the municipal *lex domicilii* that in space law has also a special dimension, *lex loci registrationis*. Unlike territorial extensions, national jurisdiction is not prohibited regarding material extensions located in the extraterrestrial realms.

[Some] asteroids and comets could be viewed not as landed extension, but as movables. According to Simpson (1976, pp. 5–6), land has two "special characteristics which distinguish it from all other commodities known to commerce", namely – "it is immovable, and so it cannot be physically transferred from one person to another", and "it is everlasting . . . [t]he owner of land . . . [cannot] in its legal sense, destroy it; his power is limited to the enjoyment or disposition of rights in or over it". [Some] asteroids and comets, however, do not have these characteristics; with the appropriate technology, they could be moved; and they can be destroyed, i.e. consumed in their totality. Thus, they may qualify as movables.

3.4 The Different Approaches in Defining Celestial Bodies

While there is still a number of authors that do not differentiate between natural bodies in outer space and thus include the asteroids and comets under the sway of the non-appropriation principle, some other authors do differentiate between celestial entities, perceiving movable/material ones as different from immovable/spatial ones. Several theories of defining what a celestial body is and what it is not, have appeared. It is to be noted that the names of the theories are not the ones given by the authors. In the quest for a solution to another unresolved space law problem – the definition of outer space as opposed to Earth's atmosphere –, two main schools of thought have formed, namely the spatialist and the functionalist one. The cognate delimitation of the territorial sea from the high seas has also been subject to a sequence of approaches. It has started with the control approach, the cannon-shot rule stating that the place where the power of the arms ends is the same as the place where the national territory ends – "*terræ potestas finitur ubi finitur armorum vis*"; or "*ubi vis ibi jus*" – where there is [en]force[ment], there is law. While at that time the range of a cannon shot was about three miles, the approach turned into a spatialist one, the breadth of the territorial sea being generally regarded as three miles even if the cannons were soon able to shoot further away. A revision of the distance did occur in some countries that extended their territorial sea at 12 miles. Nowadays, an exact delimitation between a national sea and an international sea cannot be spatially drawn, being in fact functionalist as it depends on the activities occurring, e.g. fishing, navigation, law enforcement, etc.

From this it can be seen that each approach has its own merits and served its purpose at its time.

3.4.1 The Spatialist Approach

In the similar quest for a definition of outer space as opposed to airspace, a "spatialist" school of thought has been formed – seeking to discover a spatial limit where the atmosphere would legally end and the outer space would legally begin. According to this, "air law and space law would cover the space above the Earth's surface split into two slices by different legal regimes as the legal status of territorial sea differs from that of the free open space" (Gal, 1997, p. 125). Applied to the present topic, a spatialist approach would define celestial bodies as objects over a certain size, while objects under that size would not be celestial bodies. The practical problem is to quantify that size, and to reach a consensus over that. It is far from a simple quest, falling under the spell of the "sorites paradox". Basically, some concepts are vague, lacking sharp boundaries; such is the case for a heap (*σορoσ* in Greek), as discovered by the philosopher Eubulides. In Hyde's (2005) words:

> Would you describe a single grain of wheat as a heap? No. Would you describe two grains of wheat as a heap? No. . . . You must admit the presence of a heap sooner or later, so where do you draw the line?

In our case, if we accept that the Moon is a celestial body and, on the other hand, a piece of dust floating in space is not, where does one draw the line between celestial bodies and space dust? At what dimension a "stone in space" ceases to be legally movable, and becomes legally immovable?

Difficult as it is, lawyers have the ability to find the mythical "straw that broke camel's back". Where there is no natural boundary or one cannot discover it, law can set a conventional boundary. Such is the case with the legal age – while virtually as [im]mature as a day ago, a person that just turned 18 has different rights and responsibilities than before. The legal age is not a completely arbitrary creation, and while it varies from a jurisdiction to another the differences are not very significant. However, formulating a legal limit between outer space and air space is subject to many avatars – Gyula Gal (1997, p. 126) lists 49 spatialist proposals suggested by 1964, ranging from 12 to 38,400 km and *ad infinitum.* Thus, while a legal limit can break the sorites paradox, the practical problems are to suggest a limit that should not be arbitrary, and to find a consensus over that limit.

The analogous experience of setting a spatial boundary between the territorial sea and the high seas is interesting insofar as it has evolved from a control approach that at a certain time extended to three miles. Even if the reach of possible control has expanded, the conventional spatial limit remained. Thus, should initially a control approach be applied and should it be possible thus to move asteroids as big as 100 m long, this may then impose itself as a spatialist limit even if later on the technique would permit the displacement of bigger asteroids.

Unlike the case for the delimitation of airspace from outer space, the supporters of the spatialist school did not come with 49 different proposals as illustrated *supra.* They mainly came with questions formulated in a spatialist manner.

Thus, Fasan (1980, p. 9) inquires whether a meteorite, following its natural orbit in outer space, is a celestial body or not:

> If it has got a diameter of few meters, is it then still permissible to "use" outer space by catching this meteorite and bringing it down to earth? And if not, what about a sole pebble with two centimetres of diameter or what about a few particles of dust? And if such an exemplary meteorite would be "appropriable", what then about an asteroid of the same size moving within the asteroid belt between Mars and Jupiter and, sometimes, even coming much nearer to Earth?

Brooks (1966, pp. 322–323) agrees with the supposition that "micrometeoroids in space, usually no bigger than a grain of sand, are subject to appropriation by the finder, as they have always been after falling to earth", but he admits that "[s]ome difficulty may be encountered as one moves toward objects of larger sizes. Is a medium sized asteroid a celestial body or a floating mineral resource?". He accepts that "[n]o legal impediment would seem to stand in the way of appropriating the entirety of an asteroid, though for policy reasons a state may deem this as inadvisable."

Sztucki (1966, p. 64) considers asteroids as being celestial bodies in the legal sense, but not the meteorites that are:

> celestial bodies in the astronomical sense but certainly cannot be subjected to legal regime envisaged for celestial bodies and e.g. excluded from appropriation. There is, however, an essential difference between meteorites and asteroids. Freedom of exploration and use of outer space, naturally, presupposes taking samples of meteorites, etc., which because of the unaccountable number of meteorites and no fixed trajectory, does not impair possibilities of other states to do exactly the same.

Regarding dimensions, he writes that "the smallest asteroid known yet as some 0.3 km. in diameter.

Zhukov (1967, p. 273) considers as celestial bodies in the OST legal sense the "planets and their natural satellites, asteroids, and large meteorites", but excludes "micrometeorites, smaller meteorites, and comets", the latter presenting "more grounds for their referring directly to outer space."

Williams (1969, p. 179) finds useful a definition that has been embraced at the First Colloquium on the Progress in Cosmic Exploration and its Consequences upon Humanity, held in Buenos Aires in 1966, viewing a meteoroid as "a solid object moving in outer space, of considerably smaller proportions than an asteroid but considerably larger than an atom or molecule." This is in fact a definition adopted by the IAU Commission 22 in 1961 (Point B). The problem of vagueness remains – how much is "considerably smaller" and "considerably larger"?

A spatialist flavour follows from the text of the Outer Space Treaty, that in Article I provides for the "free access to all areas of celestial bodies." From this, it results that one cannot consider a small space rock as a celestial body insofar as it is not viewed primarily as an area permitting landing on it.

The spatialist approach has its merits insofar as it distinguishes between small objects – which are not celestial bodies – and big objects, which are celestial bodies. However, the problem still remains to agree on how small is small.

3.4.2 The Control Approach

The effective control approach has been used in the law of the sea as the "cannon shot rule", and Gal (1997, p. 127) cites examples of its applicability also as one of the criteria of delimiting air space from outer space. In a slightly modified form, the control approach would distinguish between immovables – celestial bodies – and movables in outer space literally, according to the actual ability of moving them. It is an approach that stands on very logical arguments, proven elsewhere. Thus, according to Reid (1996, p. 18), "[t]raditionally, movable property is described as being property which either moves by itself or which can be moved by others", quoting Bell's Principles: "Whatever moves, or is capable of being moved from place to place without injury or change of nature in itself, or in the subject with which is connected, is movable".

While comfortable at first sight, Reid admits that, rather than a definition, this is in fact a description, as "some movable property is so substantial that in practice is never moved, while, as Stair (1693, II, I, 2) points out, the fact that soil is capable of being dug up and moved, or that 'the sea ... hath its agitation by ebbing and flowing' does not prevent both from being classified as heritable". In the same time, law came with the fiction of immovables by destination, i.e., things that physically move or can be displaced but that are regarded by law as being immovables.

While the fiction of immovables by destination pertains to municipal law, and while the range of immovables by destination is different in each system of municipal law, it cannot be applied to outer space except by means of international treaty. Thus, in outer space there can be only movables or immovables by nature.

The recourse to natural law would classify as immovables the "things that do not move", while movables would be "things that move". Nevertheless, natural law may prove tricky in this case. When extended at the cosmic scale of space law, everything moves by itself – thing admitted as early as ancient Greece, where παηθα ρει – everything flows. Smith and Zaibert (1997) agree – "[t]he distinction between movables and immovables is itself vague; there are, strictly speaking, no immovable objects". This is especially true in the extraterrestrial context. Land is the immovable *par excellence*, but it orbits the Sun at 18.5 miles/sec, while the entire Solar System orbits the center of the Milky Way at 140 miles/sec (Stardate, 1995). So, in deciding what is movable or immovable in outer space, the criteria of movability by itself cannot be applied, otherwise everything in outer space would be movable.

This is where the second thesis of the movability comes into service, namely the classification as movable of the property that can be moved through human intervention. According to the control approach, it is movable what it can actually be moved, and immovable what it cannot be moved. By the action of actually moving it, one makes it appropriable. Change occurs in the moment of actual movement; property would install when moved.

While not properly authoritative, the most relevant definition of celestial bodies follows a control approach. Most of the scholars that have studied the problem of defining celestial bodies belong to the control school. In the 1960s, the members of Working Group III of the International Institute of Space Law concerning the legal status of celestial bodies came with a definition considering celestial bodies in the legal sense as "natural objects in outer space ... which cannot be artificially moved from their natural orbits" (Smirnoff, 1966, p. 13).

It is indeed envisageable that asteroids can be moved in the near future, and law should deal with this. Nevertheless, it is not yet the time to regulate projects that, if not impossible, hold nevertheless of the domain of the very distant future; thus are the fantastic ideas of Freeman Dyson to build a sphere around the Sun from all the materials contained in the planets; Earth would be thus used in its substance. Of the domain of science-fiction holds also the idea of moving the Earth on another orbit – possibility already examined in the 1961 British movie "The Day the Earth

Caught Fire" where Earth is catastrophically displaced towards the Sun by some polar nuclear explosions, and then scientists try to put right the wrong by the same means.

The idea of moving the Earth is not so new – Archimedes is said to have offered to move the Earth with levers if given a place to stand. And in 2001 his idea has been reiterated by three US scientists – Korycansky, Laughlin, and Adams (quoted in Whitehouse, 2001), that consider that mankind will soon have the ability to move the 5.972 sextillion tonnes Earth into a new orbit, the planetary manoeuvre being envisaged to more than double the time life can survive on our planet. They envisage repositioning the Earth to maintain a benign global climate by using a large asteroid through the gravitational sling shot technique. This would counter the fact that in the next billion years the Sun will increase its brightness and if the Earth stays in its present orbit it will be fried and all life eliminated. Zwicky (1972, p. 265) imagines even the possibility of moving the Sun once the means are available near Alpha Centauri.

As with any theory, the control approach is not accepted by everybody. In July 1980, testifying in front of the US Senate, NASA General Counsel Neil Hosenball expressed his view that if an asteroid were moved into Earth orbit for exploitation, it would still be a celestial body within the meaning of that term, and would not change its character by its moving (Leich, 1980).

However, there is merit in considering that the process of actually moving an asteroid/comet would qualify as extraction, the body in cause ceasing to be a resource "in place" and thus by-passing even the general prohibition in Article 11.3 of the Moon Treaty.

3.4.3 The Functionalist Approach

According to the functionalist approach in the delimitation of airspace from outer space, space law "can not be associated with a limited space, but only with the character of the activity under regulation" (Gal, 1997, p. 126). Functional approaches exist in other areas of municipal law, which classify some movables by nature as legally immovable. In our field, a functional approach would differentiate between objects used in their spatial dimension – these being deemed as celestial bodies – or in their material dimension, these being movable orebodies; or, if used for navigation, they would be space objects. A functional approach would take into account the actual use of the asteroid – i.e. for building of a base, for exploitation of resources, or for navigation.

3.4.4 The "Space Object" Approach

As seen *supra*, some daring scientists would envisage using asteroids as interplanetary/interstellar vehicles; the legal classification of such objects as space vehicles would be supported by a functional approach. A spatialist approach would certainly be unsuitable in differentiating between space objects and celestial bodies, given that some natural bodies in space would be smaller than the intended Solar Power Satellites, planned to reach 10 × 5 km each in surface (Diederkis-Verschoor, 1980, p. 149).

Fasan (1984, p. 243) envisages the use of natural Celestial Bodies as "space objects", and examines the case when an asteroid, by way of human intervention, is reformed, used as a shell for a space station, losing its natural appearance, together with its legal status of "celestial body" by becoming a manmade structure, i.e. legally a space object. This so-called "Asteroid Base" would then have to be registered internationally with the Secretary General of the United Nations.

Another case can be put forward where [some] asteroids, even if not proper used as space objects, would be considered so instead of celestial bodies. While Benson (1998, p. 46) reckons that there is no appropriate body to which he could make his asteroidal claim, the only possible option being to make the claim to the public, should small asteroids be considered space objects they could be claimed by way of registration in the national registry of space objects referred to by Article VIII of the Outer Space Treaty:

> A State Party to the Treaty on whose registry an object launched into outer space is carried shall retain jurisdiction and control over such object.

These provisions have been detailed in the 1975 Convention on Registration of Objects Launched into Outer Space, that in Article I(c). defines the term "State of Registry" as "a launching State on whose registry a space object is carried..." and provides both for a national and international registration of space objects. An international Registry is maintained, in accordance to Article III.1, by the UN General Secretary "in which the information furnished in accordance with Article IV shall be recorded"; Article IV requires each State of Registry to furnish to the UN General-Secretary information concerning each space object carried on its registry, namely name of launching State or States, an appropriate designator of the space object or its registration number, date and territory or location of launch, basic orbital parameters, including nodal period, inclination, apogee, perigee, and general function of the space object.

The Registration Convention is very liberal insofar as it gives the State of registry concerned the freedom to determine "[t]he contents of each registry and the conditions under which it is maintained" (Article II.3). This may be interpreted as entitling a State to register small asteroids as space objects on its registry.

Should a launching State refuse the request of a private entity to register a small asteroid as a space object, a way of lawfully cheating this provision would be to borrow from the maritime law the institution of "flags of convenience" and to find a State willing to assume the position of launching State and to register the small asteroid. The Registration Convention provides for the situation when there are two or more launching States, according to Article II.2 these needing to "jointly determine which one of them shall register the object". A conflict could thus arise between an unwilling actual launching State, and a willing "convenience" State.

Of course, the status of launching state would come with its privileges and responsibilities, one of the latter being its international liability as provided by several Space treaties. In the light of this international liability, the registration of small asteroids as space objects and their private ownership would be in fact of benefit to the possible survivors of accidents provoked by small asteroids. Several questions spring from this possible approach – such as whether compulsory insurance should be legally imposed upon owners – like the one imposed to atomic energy providers. Would this discourage people in claiming asteroids? Were asteroids *res nullius*, nobody would be responsible for damage caused by them; were they *res communis*, then the whole humankind would be responsible for the damage to a private spaceship or other object by such a common asset.

An objection that may be brought is that there cannot be a launching state in the case of asteroids, as these exist already in outer space and no launching occurs. The Outer Space Treaty provides however in Article VIII for the possibility of space objects not being launched from Earth, but being instead manufactured in outer space, speaking about "objects launched into outer space, including objects . . . constructed on a celestial body . . . found beyond the limits of the State Party to the Treaty on whose registry they are carried". From this it is understood that objects constructed on a celestial body, where no actual launch occurs, are included in the category of objects launched into outer space; furthermore, these need to be carried on the registry.

3.4.5 The Iceberg Analogy

Interesting consequences would result from using the legal status of icebergs as a paradigm for the legal classification of comets and asteroids. In fact, comets have been often described as "dirty snowballs" and the "icebergs of space", offering striking similarities to icebergs (composition, dimension, location in an international area). Indeed, Comet Wirtanen is a ball of rock and ice just 600 m across, which, if gently landed on an Earth ocean, might be legally considered an iceberg. It is to be noted that icebergs have as well a rather unclear legal status, though their small-scale exploitation has already begun. Like asteroids and comets, icebergs have a spatial dimension but are used mainly in their material dimension, as a floating mineral resource. While Article 89 of the UN Convention on the Law of the Sea prohibits the national appropriation of the high seas, we have no knowledge of States having protested appropriation of icebergs. In the same time, we have no knowledge of a formal declaration of ownership over icebergs by the entities using them in their material extension; the principle of extraction seems to apply, given that icebergs have been appropriated either in their entirety and displaced from their initial location, or parts of them have been moved away without claims being laid for the exclusion of others from the exploitation of that particular iceberg.

3.5 Conclusion

Given that, as shown *supra*, the notion of celestial bodies is not legally defined, it may be possible that several asteroids might escape the non-appropriation principle because of this lacuna. The issue of defining celestial bodies is extremely intricate, and there is no absolute answer to be given *ex cathedra*. We have only attempted to present the existing theories and some new approaches, but at the end of the day only practice will decide whether [some] asteroids are places or movables.

Chapter 4
The Relationship Between Property and Sovereignty in Outer Space[1]

-You own the stars?
-Yes
-But I've already seen a king that...
-The kings do not own. They "rule" over. It is very different.
Antoine de Saint-Exupéry (1946)

4.1 Introduction

This chapter will analyse the relationship between appropriation under international law, and civil law appropriation. It will try to answer the age-old question: are sovereignty and property, as State-powers over land and private-powers over land, independent concepts, or they are interlinked? Smith and Zaibert (1997) recognise that "[i]t is not always easy to distinguish between these powers, insofar as they have traditionally been viewed as belonging to the subject-matters of separate disciplines". "Property and Sovereignty", considers Morris Cohen (quoted in Smith and Zaibert, 1997), "as every student knows, belong to entirely different branches of the law. Sovereignty is a concept of political or public law and property belongs to civil and private law".

This chapter will examine, in this particular case, whether the non-appropriation principle on the international plane, as contained in the Outer Space Treaty, results also in the prohibition of the appropriation of the celestial bodies on the plane of private property rights. It is to be noted that at this point only the implications over land and natural resources in place will be examined – property rights over planetary resources removed will be treated separately as subject to different regulations.

4.2 The Prohibition of National Appropriation in Outer Space

As defined by Swiss arbitrator Max Huber in the *Island of Palma Case*, territorial sovereignty – a concept synonymous with independence – is "the right to exercise

[1] A version of this chapter has been published as Pop, V., 2000. Appropriation in outer space: the relationship between land ownership and sovereignty on the celestial bodies. *Space Policy*, Vol. 16, p. 275.

V. Pop, *Who Owns the Moon?*, Space Regulations Library Series 4,
DOI 10.1007/978-1-4020-9135-3_4, © Springer Science+Business Media B.V. 2009

therein, to the exclusion of any other State, the functions of a State". As we have shown *supra*, one of the functions of a State is to regulate landed property rights within its own territory. Consequently, in order to find out who owns the Moon, we have to find out which State is the sovereign on the Moon. The answer is to be found in the 1967 Outer Space Treaty, whose Article II clearly states:

> Outer space, including the moon and other celestial bodies, is not subject to national appropriation by claim of sovereignty, by means of use or occupation, or by any other means.

In other words, no State Party to the Outer Space Treaty can lawfully obtain sovereignty on the Moon and other extraterrestrial realms. This non-appropriation principle is one of the back bones of contemporary international space law, contrasting sharply with the old school of *terra nullius*, yet being in harmony with the doctrine which followed the end of World War II. But why is the *lex lata* adverse to national appropriation in outer space? What caused the law to be as it is – and, more important – are these causes still of actuality?

Several factors have shaped the non appropriation principle of the Outer Space Treaty. The first reason for the rejection of territorial sovereignty in outer space was the *a priori* avoidance of conflict – a central theme in the post-bellum international law. The signers of the 1945 United Nations Charter agreed upon its text with the determination to "save succeeding generations from the scourge of war" (Preamble), and entrusted the newly created UN, *inter alia*, with the task of "removal of threats to the peace" (Article 1.1). Areas that would have been normally considered *terra nullius* (the Moon, the deep seabed) or where conflicting claims could have grown into full-fledged wars (Antarctica) became, over the years, objects of international treaties. Often, the prevention of conflict was mentioned *expressis verbis* in the text of the international document as a material source. The signatories of the 1959 Antarctic Treaty reached agreement while "[r]ecognizing that it is in the interest of all mankind that Antarctica ... shall not become the scene or object of international discord" (Preamble), whereas the States Parties to the 1979 Moon Agreement have agreed upon its text "[d]esiring to prevent the moon from becoming an area of international conflict" (Preamble). Darrel Menthe (1998, para. 47) agrees that "[i]n each international space, the specter of international conflict has been a prime factor in forming treaty regimes", whereas Wayne White (1985, p. 16) recalls the conviction of most policy analysts that "territorial claims and counterclaims – the extension of national rivalries into outer space – would inevitably result in some form of international conflict". Not only politicians, but also voices from the public, wanted to avoid hostilities in outer space. "Twenty-five years hence" – asked a 1953 editorial – "will there be British Luna, American Luna, Soviet Luna, Indian Luna, French Luna? Will warriors in space suits battle across the silent airless wastes for possession in the names of their nations? Or can we settle this thing now while an objective viewpoint is possible?" (The Chronicle Telegram, 1953, p. 23). In November 1958, the statement of the Australian delegate in the United Nations First Committee (quoted in Jessup and Taubenfeld, 1959, p. 274) echoed this concern:

> Experience in Antarctica may suggest how difficult it may become to consider the problems of outer space impartially and on a universal plane if decision is left until states have established themselves permanently in the field.

Another argument for the acceptance of the non-appropriation principle is cosmetic; a different view, supporting territorial sovereignty in space, would be "politically incorrect" and ruin one's prestige and influence. According to White (1985, p. 18), Soviet and American space policy was highly politicized in the period between 1957 and 1969, when the two superpowers were competing for the allegiance of the newly independent African and Asian countries. As the latter countries were opposed to imperialism in all its forms, both USSR and the US counted on an enhanced prestige and political influence by rejecting "territorial sovereignty and its overtones of colonialism". Menthe (1998, para. 43) concurs that, as it is the general case with international spaces (deep sea bed, Antarctica, the moon), former colonies identified *res nullius* with imperialism – hence opposed the old doctrine enshrined in Roman law and in the Lockean model of natural law.

White (1985, pp. 17–18) brings forth two other reasons behind the non-appropriation principle: to warrant free access to all areas of outer space, as it has been the case with the high seas, and to avoid the technical difficulties that would be encountered in establishing title and delineating boundaries in outer space. Yet an interesting hypothesis has been advanced by Robert Zubrin (1999, pp. 12–13), namely that the authors of the Outer Space Treaty advanced its principles in order to "remove space from the highly charged domain of cold war competition, thereby allowing the space program to be shut down in order to make its funding available for other projects". Zubrin supports his view with a letter from US Assistant Secretary of State Henry Owen to National Security Advisor Walt Whitman Rostow dated December 9, 1966 – letter obtained by Alan Wasser under the Freedom of Information Act. In this letter, Owen advocated an ending to the space race – and the Outer Space Treaty would provide for such an ending – because "[i]t will save money, which can go to (i) foreign aid, (ii) domestic purposes, – thus mitigating the political strains of the war in Vietnam". According to Zubrin (1999, p. 14), this strategy worked: "Within two years of the treaty's ratification in 1967, US space funding dropped by 26 percent. Within four years it was down 45 percent. Within six years it was down 60 percent".

We will discuss later on whether the reasons that shaped the law half a century ago are still valid today. What matters, for now, is that, *de lege lata*, no State can lawfully appropriate the extraterrestrial realms or parts thereof.

4.3 What Is Property?

In our attempt to find out who owns the Moon, we have to elucidate what "owning" means. "There is no image, no painting, no visible trait, which can express the relation that constitutes property" – wrote English jurist Jeremy Bentham (1802, Chapter VIII). Property – said he – "is not material, it is metaphysical; it is a mere conception of the mind". Two centuries later, property still escapes the grasp of legal scholars. "Few other legal notions operate such gross or systematic deceptions" – considers Kevin Gray (1991, p. 252):

> With private property, as with many illusions, we are ... seduced into believing that we have found an objective reality which embodies our intuitions and needs. But then, just as the desired object comes finally within reach, just as the notion of property seems reassuringly three-dimensional, the phantom figure dances away through our fingers and dissolves into a formless void.

Is property in outer space as vacant and formless as the void thereon? The reason property escapes through one's fingers is not its ghostly form; it is instead its loose nature. Far from being a monolithic construct, property is in fact a bundle of rights.

According to Neil Meyer (2000), property represents in fact "the access right to a stream of benefits from a given set of resources." The view of ownership as a modular concept carries on from Roman law, where *dominium* was seen as an embodiment of three attributes – "*jus utendi, fruendi et abutendi re sua quatenus juris ratio patitur*" – i.e., "the right to use, to enjoy the fruits and to abuse one's own good insofar as law allows this". It is, therefore, as flawed to discuss about whether private property rights exist in outer space as it is attempting to catch a school of fish with a hook.

Land ownership is a multi-dimensional concept. Far from being absolute, property is bound in space by the physical limits of the plot of land; it is bound in time by the duration of the owner's life; and it is bound in law by the limits imposed by the society. As seen *supra*, law is not a static phenomenon. The concept of property rights – as many legal relationships – has changed over time due to the transformations in the society, and the bundle of rights attached to land ownership is different today from what it was a decade ago (Acland, 2003, p. 30). Property is, therefore, a variable concept.

We have been inspired by the "Rousseau-Hobbes" theory, according to which "real estate arises when raw land is fenced and these fences are secured and policed via State or kingly power" (Smith and Zaibert, 1997). This is very much in line with the meaning attached to property by Kevin Gray (1991, p. 299). His view is that property is a "legally endorsed concentration of power over things and resources". He recognises that "the State takes on a critical, and so far little explored, role in defining the concept of 'property'", becoming "a vital factor in the 'property' equation". Still, the role of the State in this equation has been "much overlooked in the traditional common law accounts of private property" (Gray, 1991, p. 304).

4.4 The Impact of the Non-Appropriation Principle Over Property Rights

The role of the State has been addressed by the Outer Space Treaty. While Article II forbids *expressis verbis* the "national appropriation by claims of sovereignty, by means of use or occupation, or by any other means" of outer space, including the moon and other celestial bodies, no explicit mention is made regarding their private appropriation. The language of the Moon Agreement of 1979 is more precise, providing that:

> [n]either the surface nor the subsurface of the moon, nor any part thereof or natural resources in place, shall become property of any State, international intergovernmental or non-governmental organisation, national organisation or non-governmental entity or of any natural person (Article 11.3).

The Moon Agreement has a very poor ratification record and, as *pacta tertiis nec nocent nec prosunt*, it cannot serve as a standard for the whole of the international community. A thorough analysis of the widely accepted Outer Space Treaty is therefore needed.

The provisions of Article II of the Outer Space Treaty have received different interpretations from legal commentators that will be further examined.

4.4.1 First Viewpoint: Outer Space Treaty Allows Private Appropriation

A minority of authors consider that the Outer Space Treaty prohibits only the national appropriation of outer space and celestial bodies; *per a contrario*, appropriation by other entities is allowed. This is the view held by Gorove (1968, p. 42), when writing:

> While further developments in space law, by international custom or treaty, may eventually prohibit spatial appropriation by John Doe ..., the [Outer Space] Treaty in its present form appears to contain no prohibition regarding individual appropriation .... Thus, at present an individual acting on his own behalf or on behalf of another individual or a private association or an international organisation could lawfully appropriate any part of outer space, including the moon and other celestial bodies.

The role of John Doe has been taken by the many dealers in "unreal estate" mentioned in the first chapter. As detailed *supra*, Dennis Hope's claim to the Moon and other celestial bodies is partly based on the assumption that the Outer Space Treaty contains a "loophole" allowing the private appropriation of the extraterrestrial realms. In his logic, while the Outer Space Treaty forbids *expressis verbis* any government from claiming a celestial body, the document does not mention individuals or corporations. This "loophole theory" seems to have been swallowed by a part of the press, charmed by the impression that Hope has "outsmarted" the UN. Even a contemporary legal commentator believes that "[i]t's smart reasoning. There is no actual specific clause in any legal document to say Mr Hope is wrong" (Frans von der Dunk, quoted by Dixon, 1998, p. 29).

We reject the "loophole" theory as being extremely narrow; the arguments for doing so will be further presented.

4.4.2 Second Viewpoint: Outer Space Treaty Prohibits Private Appropriation

The second view, shared by most of the authors, holds that private property is denied under the Outer Space Treaty (O'Donnell and Goldman, 1997, p. 322). Several methods of legal reasoning sustain their view. Application of the *a fortiori* principle to the prohibition of national appropriation of outer space and celestial bodies, as expressed in the Outer Space Treaty, results in its implicit extension to private parties. Sterns et al. (1996, p. 53) hold that States may not license private entities to "appropriate privately that which cannot be appropriated publicly". C. Wilfred Jenks (1965, p. 201) believes that, as "States bear international responsibility for national activities in space; it follows that what is forbidden to a State is not permitted to a chartered company created by a State or to one of its nationals acting as a private adventurer".

Historical interpretation would sustain the same viewpoint. Prior to the negotiation of the Outer Space Treaty, many legal institutes viewed the private appropriation of extraterrestrial realms as undesirable. In 1960, the Association of the Bar of the City of New York recommended that "Celestial bodies ... shall not be subject to exclusive appropriation by any person, organisation, or State on the Earth" (Provision U). Three years later, the Institute of International Law suggested that "Outer space and the celestial bodies are not subject to any kind of appropriation ..." (Principle 1). The same was the view of the International Institute of Space Law in 1966, whose Working Group III drafted a resolution concerning the legal status of celestial bodies; its third principle held that "[c]elestial bodies or regions of them shall not be subject to national or private appropriation ...".

The negotiating history of the Outer Space Treaty itself points towards the implicit prohibition of private appropriation. Dembling (1997, p. 39) considers that the text of Article II "provoked little debate". Towards the end of the Outer Space Treaty negotiations, on August 4, 1966, Delegate Bal of Belgium (quoted in Oosterlinck, 1996, p. 274) stated that his delegation:

> had taken note of the interpretation of the term 'non-appropriation' advanced by several delegations – apparently without contradiction – as covering both the establishment of sovereignty and the creation of titles to property in private law.

The French Delegate also mentioned that:

> ... there was reason to be satisfied that three basic principles were affirmed, namely: the prohibition of any claim of sovereignty or property rights in space... (quoted in Oosterlinck, 1996, p. 274).

Opposing views, such as that of Oosterlinck (1996, p. 274) consider however that the history of Article II, as recorded during the negotiating meetings, shows that:

> it was the intention of at least some participants to give a wider scope to this article than its actual reading.

Eugene Brooks also comments that:

> [i]t would be more correct to say that several delegations had raised, rather than settled, the precise point (Brooks, 1968, note 23 at p. 350).

and that:

> the entire record is more evidence that other contracting parties did not wish to foreclose future positions by setting forth precisely the rights of the Powers in using the resources of celestial bodies. In that sense the question may still be open (Brooks, 1968, p. 344).

However, the fact that the accessions to the Outer Space Treaty were not accompanied by declarations expressing the understandings of the respective nations regarding the meaning of Article II may be seen as an indication of the fact that the matter was settled during the negotiation process.

Reductio ad absurdum reasoning shows that, should private appropriation be permitted, this may be converted by means of the doctrine of eminent domain into State property. The US government would be entitled to confiscate the lunar real estate of the US citizen Dennis Hope for public use, thus making the Moon become US public property. Nevertheless, national appropriation of the extraterrestrial realm is forbidden by the Outer Space Treaty phrase "by any ... means". Application of the doctrine of eminent domain would be one of these indirect means.

The teleological interpretation of the Outer Space Treaty examines the ultimate aim of the instrument. According to David S. Myers (1975, p. 68), the Outer Space Treaty has conferred upon outer space the character of "*res communis* – placed by nature at the equal disposal of all men and non-appropriable by individual States or private persons".

The systematic method of interpretation relates the norm in issue to the whole of the legal text. Property implies control over access; its essence is the exclusion of the others for the benefit of the owner. As Article I of the Outer Space Treaty states that "there shall be free access to all areas of celestial bodies", no control over access can be lawful, hence private appropriation of extraterrestrial land cannot exist.

Finally, another reason that may be invoked when affirming that prohibition of national appropriation implies prohibition of private appropriation is the position according to which private appropriation cannot exist independently from State appropriation. When appropriating a previously unoccupied land, one does so necessarily on behalf of a State. Lauterpacht (1955, p. 555) believes that "occupation can only take place by and for a State; it must be a State act, that is, it must be performed in the service of a State, or it must be acknowledged by a State after its performance". Unless the State invests a private individual or corporation with the public power of acquisition and administration:

> acquisition of territory and sovereignty thereon takes place outside the dominion of the Law of Nations, and the rules of this law, therefore, cannot be applied. If the individual or corporation which has made the acquisition requires protection, he or it must either declare a new State to be in existence and ask for its recognition by the Powers ..., or must ask an existing State to acknowledge the acquisition as having been made on its behalf (Lauterpacht, 1955, pp. 544–545).

We will comment in the next section upon the possibility of individuals to acquire landed property outside State sovereignty.

4.4.3 Third Viewpoint: Private Appropriation Is Not Legally Enforceable

It is hereby offered that, regardless whether prohibited or not, landed private property cannot survive outside the sphere of sovereignty. What can exist is a mere factual situation that will or will not be subsequently converted into property by means of State endorsement. As sovereignty is outlawed by the Outer Space Treaty, such a State endorsement cannot occur in regard to the private occupation of land on celestial bodies.

Indeed, in order to exist, property needs a superior authority to enforce it, be it in the form of a State or some other recognised entity. Stubkjaer (quoted in Smith and Zaibert, 1997) considers that "[w]ithout a society a person would hold land in possession rather than own it". A *de facto* appropriation can occur, yet property *per se* cannot exist outside the sphere of State protection. As sovereignty or sovereign rights are outlawed, landed private property on the celestial bodies is crippled due to the fact that it is not enforceable, such enforcement being deemed a violation of the non-appropriation principle.

Besides philosophical arguments, our views are supported by legal precedents. Two years before the placing of the island of Jan Mayen under Norwegian sovereignty by means of a Royal Decree of May 8, 1929 (Hackworth, 1940, p. 475), the US Department of State exchanged correspondence with Hagbard Ekerold, a US citizen and owner of the Polarfront Company, a New York Corporation regarding the Company's claims to rights on the island, at that time still *terra nullius*. The US Department of State expressed its view that "the general recognition of the status of the island as *terra nullius* rendered it impossible to acquire title to property there, as ordinarily understood". In the words of the Department:

> Ownership, in its essential features, constitutes the use and enjoyment of the property owned, to the exclusion of all others in its use and enjoyment, and is secured to the owner under the authority of the Government exercising the right of sovereignty with relation both to the island and its inhabitants (Hackworth, 1940, p. 476).

Subsequent to the establishment of Norwegian sovereignty over the Jan Mayen island, the US Department of State informed the Norwegian Minister in Washington about the fact that the US Government is "confident that the Norwegian Government will not fail to respect the rights of Mr. Hagbard D. I. Ekerold and the Polarfront Company". The Norwegian legation shortly replied that "the occupation of Jan Mayen by Norway was in no way intended to cause changes in the rights which, according to civil law, 'exist on the island'" (Hackworth, 1940, pp. 475–476).

In a related case before the Supreme Court of Norway – *Jacobsen v. Norwegian Government* (1933) – , the Court recognised the plaintiff's entitlement to undertake a private occupation in Jan Mayen with the object of obtaining property in the occupied land, while the Norwegian Government was not recognised its entitlement to proprietary rights in the part of the island which had been occupied by the plaintiff.

A similar situation existed on Spitzbergen. Until 1920, when Norwegian sovereignty over the archipelago was recognised by means of an international treaty, the land in issue was considered to be *terra nullius*. Several private entities of various nationalities occupied land in Spitzbergen, notifying their claims to land to their respective governments. No government asserted sovereignty over the Spitzbergen on the basis of the occupation performed by their nationals, but national "interests" arose. US President Taft, addressing the US Congress on 3 December 1912, stated that:

> The great preponderance of American material interests in the subarctic island of Spitzbergen, which has always been regarded politically as "no man's land", impels this Government to a continued and lively interest in the international dispositions to be made for the political governance and administration of that region (Lauterpacht, 1955, pp. 465–466).

The Spitzbergen situation was recognised by Robert Lansing (1917, p. 764) as a "unique international problem":

> If this [mining] population increased and persons of different nationalities settled in the islands laying claims to lands already claimed by others, how would these people be governed and to what authority could they appeal to settle their conflicting claims and to protect them in the enjoyment of their rights? ... [T]he fact that the principle of *terra nullius* must be considered as a factor it is by no means easy of solution.

The 1920 Spitzbergen Treaty settled this problem by recognising the "acquired rights of nationals of the High Contracting Parties" (Article 6), and providing for an Annex to deal with "[c]laims arising from taking possession or from occupation of land before the signature of the present Treaty" (Article 6). According to the Annex, notifications of all claims to land which had been made to any Government before the signature of the Spitzbergen Treaty were to be sent by the Government of the claimant, within three months from the entry into force of the Treaty, to an independent Commissioner nominated *ad hoc* to examine such claims (para. 1 clause 1). The Norwegian Government was to "take the necessary steps to confer upon claimants whose claims have been recognised by the Commissioner a valid title securing to them the exclusive property in the land in question ... subject to the mining regulations" (para. 1 clause 9). More interestingly, the Annex provides in its second paragraph for arbitration procedures regarding the claims that were disputed (para. 2), the Governments whose nationals had been recognized being entitled to appoint an arbitrator (para. 2 clause 1). The third paragraph of the Annex provided for the extinguishing of claims not settled according to the previous paragraphs.

What the Jan Mayen and Spitzbergen examples have in common is that eventually a State sovereignty was superimposed over the private appropriation of the real estate. In the same time, all these facts have taken place in *terra nullius*, whose appropriation was not prohibited. What distinguishes the Spitzbergen and Jan Mayen situations from similar facts that may occur in outer space and celestial bodies is that, unlike in the case of *terra nullius*, the national appropriation of the extraterrestrial realms is prohibited. Once the State as provider of "legal endorsement" is eliminated from the equation, possession becomes "exposed to the elements". On celestial bodies, landed possession is,*de lege lata*, denied the protective shell of the State and is left on its own. According to Kenneth Silber (1998), the Outer Space Treaty "makes no mention of private property, but it undercuts the ability of any government to recognise or enforce a private claim". Should a State recognise, endorse or protect in any way the territorial acquisition of any of its subjects, this would constitute a national appropriation in violation of the Outer Space Treaty. Maintaining an extraterrestrial land registry would fall into this category as a manifestation of State sovereignty. In the 1953 *Minquiers and Ecrehos* case between the United Kingdom and France, the International Court of Justice came to the conclusion that "Jersey authorities have in several ways exercised ordinary local administration in respect of the Minquiers during a long period of time", one of the manifestations of British sovereignty over the Minquiers mentioned by the Court being the registration, in the public registry of deeds of Jersey, of contracts of sale relating to real property on the Minquiers.

If confronted with a situation of private appropriation of extraterrestrial real estate or of resources in place, will States chose to respect the Outer Space Treaty and refuse to legally endorse the fact? The US Department of State was confronted with two such situations of appropriation of the seabed outside the limits of the national jurisdiction. While the first situation to be quoted is trivial, it is worth mentioning, due to its stunning similarities with the "Lunar Embassy" situation.

In 1944, A. Byron Hunicke informed the US Secretary of State Hull by a series of letters that, having developed means for making the ocean floor accessible to human exploitation, this would afford him "entry upon and occupancy of lands heretofore inaccessible beneath the seas". As a consequence, he claimed "full title and dominion in submerged lands and the waters above them lying seaward of the off-shore limit of territorial waters of nations abutting the sea". In order to perpetuate ownership under his claims and "to provide order where no national government can be permitted to function", Mr. Hunicke established the "Under Seas Proprietorship" and the "Under Seas Dominion". His assertion went so far as to claim that no other parties except the members of the Under Seas Proprietorship could dispose of the "extra territorial submerged lands". In its replies, the Department of State stressed that "private individuals or concerns" cannot "acquire for themselves dominion over lands beneath the high seas" and that an assertion such as that in Mr. Hunicke's letter "by a private person of ownership and dominion over submerged lands beyond the three mile limit" (i.e., beyond the limits of the national jurisdiction) has "no legal effect and confers upon such person no rights with respect to such submerged lands" (Whiteman, 1965, p. 740)

The second situation, of a more serious nature, arose in 1974, when Deepsea Ventures, Inc. filed with US Secretary of State Kissinger a "Notice of Discovery and Claim of Exclusive Mining Rights, and Request for Diplomatic Protection". Deepsea Ventures, having taken possession of a deposit of seabed manganese nodules located beyond the limits of national jurisdiction, asserted:

> the exclusive rights to develop, evaluate and mine the Deposit and to take, use, and sell all of the manganese nodules in, and the minerals and metals derived, therefrom.

Deepsea also requested:

> the diplomatic protection of the United States Government with respect to the exclusive mining rights described and asserted (Deepsea Ventures, 1974, pp. 53–54).

In reply, the US Department of US Department of State (1974, p. 66) refused to:

> grant or recognise exclusive mining rights to the mineral resources of an area of the seabed beyond the limits of national jurisdiction.

The Embassy of Canada at Washington (1974) conveyed a similar message, the Canadian Government refusing to:

> accept the assertion by Deepsea Ventures, Inc., that it has exclusive mining rights or some priority in time over that portion of the international seabed area as described in the notice to the Secretary of State, or that it has acquired any rights to that area or the resources thereof through its activities.

What Deepsea requested was a claim of exclusive access over an area located beyond the reach of the US territorial jurisdiction, and a claim over the natural resources in place. The United States refused to assert what in fact would have been territorial jurisdiction over an area on the deep seabed. Instead, in 1980, the US Congress passed the "Deep Seabed Hard Mineral Resources Act", where it provided for the conduct of US entities in respect to the resources of the deep seabed. Section 3. a contains a "Disclaimer of Extraterritorial Sovereignty", where it is stated that:

> By the enactment of this Act, the United States-
> (1) exercises its jurisdiction over United States citizens and vessels, and foreign persons and vessels otherwise subject to its jurisdiction, in the exercise of the high seas freedom to engage in exploration for, and commercial recovery of, hard mineral resources of the deep seabed in accordance with generally accepted principles of international law recognised by the United States; but
> (2) does not thereby assert sovereignty or sovereign or exclusive rights or jurisdiction over, or the ownership of, any areas or resources in the deep seabed.

As it can be seen, States are free to regulate the conduct of their nationals in relation to resources located beyond the limits of their territorial jurisdiction, but cannot regulate in respect to the resources themselves. The "conduct" is a dynamic, non-territorial category ("to do"); resources, on the other hand, are a static, territorial category ("to be"). The object of jurisdiction is in this case the conduct of nationals.

Applied Space Resources, a US company that intended to launch a lunar sample return mission, has chosen to avoid a situation similar to the Deepsea Ventures one. The company has declared, in reply to the "media coverage . . . about the possibility of private companies making territorial claims on bodies on which they land robotic craft", that "Applied Space Resources does not intend to claim territorial rights as part of the *Lunar Retriever I* mission" (Applied Space Resources, 1999).

4.4.4 Fourth Viewpoint: Private Appropriation Can Occur Under Individual Sovereignty

The fourth viewpoint builds upon a different view regarding property. According to Herschel I. Grossman (2004, p. 2), the existence of a legal system and a state is not necessary for the existence of effective property rights. While in the "Rousseau-

Hobbes" theory "real estate is ontologically dependent on police and sovereign power", a more liberal conception deems that:

> real estate is a function of trust, mutual respect, existence of neighbourly habits, and similar informal social institutions, [which is] ontologically dependent on certain customs and habits which might have existed even prior to the existence of the State (Smith and Zaibert, 1997).

Real estate on celestial bodies would not need thus the "forbidden fruit" represented by the recognition and protection of the State; the endorsing "entity" would be instead mutual respect from the other owners, whereas sovereignty would be vested in the individual itself. In Hans-Hermann Hoppe's (2001) view, every State is deficient on the economical and ethical plan; he believes that a different social order, free of the shortcomings of any form of State, is a possibility:

> The term adopted here for a social system free of monopoly and taxation is 'natural order.' Other names used elsewhere or by others to refer to the same thing include 'ordered anarchy,' 'private property anarchism,' 'anarcho-capitalism,' 'autogovernment,' 'private law society,' and 'pure capitalism'.

Jack Hirshleifer (quoted in Grossman 2004, p. 2) deems that "Anarchy...is not chaos but rather a spontaneous order." Whereas all anarchists condemn the State, anarcho-capitalists believe in property rights, regarding free people as being able of upholding their own property rights, individually or in cooperation. Instead of being a guarantor of property rights, the State is seen as an obstacle which nationalizes the matter of solving property disputes, which steals from those whose property it claims to protect, and which defines property rights very weakly and badly (Micklethwait, 1981).

Despite the generosity of this view, real human society lacks unfortunately the required mutual respect and trust. Garret Hardin (1968) explained, in his "Tragedy of the Commons", that unrestricted access to a commons results in its overexploitation due to the selfish behaviour of individuals. Hardin is justified by Max Stirner's (1844) egoism – a branch of individualist anarchism – whereby might equates with right, and possession coincides with property:

> I do not step shyly back from your property, but look upon it always as my property, in which I need to 'respect' nothing. Pray do the like with what you call my property!

Half a century ago, Philip C. Jessup and Howard J. Taubenfeld (1959, p. 276) warned that the renunciation by nation states of potential sovereignty would be consistent with a political void in the extraterrestrial realms, whilst "[w]ithout a political order there might well develop a state of competitive anarchy". The consensual character of space law, the lack of State sovereignty in the celestial realms, coupled with the existence of several attributes of property rights, result precisely in an anarchist current state of affairs on the celestial bodies – a "voluntary order". The required respect for the interests of other States parties in using the Moon, as well as the consultations provided as a solution should the same not be respected (Article IX OST) support the above view.

The current anarchical state of affairs may change, like in the case of Spitzbergen, through the establishment of a State sovereignty – be it through the abrogation of the non-appropriation principle, through the remote transition to self determination of planetary settlements, or through an international solution, as envisaged by Jessup and Taubenfeld (1959, p. 276):

> [A]n international solution implies a renunciation of potential sovereignty by national States and an assumption of some responsibility for the maintenance of order by some general international organization.

4.4.5 Fifth Viewpoint: Private Appropriation May Occur Under International Sovereignty

Whereas Article II of the Outer Space Treaty, para. 3 of the 1963 "Declaration of Legal Principles Governing the Activities of States in the Exploration and Use of Outer Space" and Article 11.2 of the Moon Agreement brand the extraterrestrial realms as "not subject to national appropriation", no mention is made to international appropriation. The above series of norms has an ambiguous meaning as to the possibility of international sovereignty on the Moon.

As early as 1959, the United Nations *Ad Hoc* COPUOS hinted that "some form of international administration over celestial bodies might be adopted" (Dembling, 1997, p. 35). C. Wilfred Jenks (1965, p. 201) believes that territory "might be appropriated by the United Nations acting on behalf of the world community as a whole", and that such appropriation is the only type not forbidden by the 1963 Declaration. Nonetheless, Jenks does not vest *de lege lata* sovereignty in the United Nations, but considers it "an open one for the future". A *lex ferenda* proposal advanced by Carlos Betancourt (1997, p. 309) calls for amending the Outer Space Treaty by providing for the sovereignty and jurisdiction over celestial bodies and natural space resources to be exercised by mankind.

The Moon Agreement does address the issue of international lunar ownership/ sovereignty in Article 11.3, providing:

> Neither the surface nor the subsurface of the moon, nor any part thereof or natural resources in place, shall become property of any ... international intergovernmental ... organization... The foregoing provisions are without prejudice to the international regime referred to in paragraph 5 of this article.

The first part of the above norm would outlaw the appropriation of the Moon by any international intergovernmental organisation – be it the United Nations. Nonetheless, the prohibition is circumvented by not impeding upon the international regime envisaged in Article 11.5 of the MA for the exploitation of the lunar resources.

De lege lata, no such regime has yet been set up, nor have the United Nations or a different global organization formally establish sovereignty on the Moon. Whereas in 1969 the Apollo 11 crew "came [to the Moon] in peace for all mankind", no United Nations flag was implanted in the lunar soil – be it as a symbolic gesture – despite the matter having been pondered within NASA (Platoff, 1992). However, in 1997, the United Nations Secretary-General Kofi Annan stated that the historic mission of the United Nations is, inter alia, "to convey to future generations the material and cultural heritage that we hold in trust for them" (Annan, 1997, para. 1). His failed proposal to reconstitute of the UN Trusteeship Council as "the forum through which Member States exercise their collective trusteeship for the ... common areas such as ... outer space" (Annan, 1997, para. 85) as well as the Common Heritage tenets of the Moon Agreement allude to sovereign or quasi-sovereign ambitions by the UN. In the cognate matter of the deep seabed, such a move was met by vocal criticism from commentators like Oliver North (2005), who stated that the 1982 United Nations Convention on the Law of the Sea "puts 70 percent of the earth's surface under the despot-doting, corrupt and unaccountable 'governance' of the United Nations". Wayne White (1985, footnote 378 at p. 81) quotes David Caronhas' view that "the seabed proposals of UNCLOS III set a precedent regarding systems of governance that is contrary to basic American interests and beliefs".

A sovereignty vested in the United Nations would clarify both the "international law" and "civil law" situation of the Moon as a territory. As the public trustee of the Moon and other celestial bodies, the UN would manage the extraterrestrial realms and, in the view of the active character of the trust, could privatize the extraterrestrial realms and could protect the landed property rights thereon. This tender is however unlikely, given the historical and ideological evolution of the UN. In the same time, a sovereignty vested in the United Nations would transcend its current powers, transforming it into a world government.

4.5 Conclusion

Appropriation of land can exist outside the sphere of sovereignty, but its survival is dependent upon endorsement from a sovereign entity. As the Outer Space Treaty prohibits the national appropriation of outer space and celestial bodies, a State endorsement would be interpreted as a means of national appropriation – hence it would be unlawful *de lege lata*. An assumption of sovereignty by the United Nations would also be unlawful, transcending its powers assigned by the UN Charter.

Chapter 5
The Commons Regime: Everybody's and Nobody's

Et quidem naturali jure communia sunt omnium haec, aer, aqua profundus, et mare et per hoc littora maris
Institutes of Justinian, II.,I.,1.

I own a flower that I water every day. I own three volcanoes that I clean every week ... The fact that I own them, is useful to [them]. But, you are not useful to the stars...
Antoine de Saint-Exupéry (1946)

5.1 Introduction

Article I of the Outer Space Treaty proclaims that the extraterrestrial realms – outer space, including the Moon and other celestial bodies – shall be "free for exploration and use by all States without discrimination of any kind, on a basis of equality", and that "there shall be free access to all areas of celestial bodies". The exploration and use of the extraterrestrial realms is declared as being the "province of all mankind". This norm effectively establishes among the States Parties an open access and free use regime on the Moon, making it a public good whose owner is everybody and nobody. Several other tenets of the Outer Space Treaty confirm and elaborate the above regime.

5.2 The Extraterrestrial Realms as a Commons

About 15 centuries ago, the *Corpus Juris Civilis* proclaimed that access to the seashore pertains to everybody on the basis that, "[b]y natural law itself these things are the common property of all: air, running water, the sea, and with it the shores of the sea" (Institutes of Justinian, II.,I.,1.). A few paragraphs later (II.,I.,5), the same norm stated that "seashores are regarded as the property of no one but being of the same legal status as the sea itself and as things lying under the sea, earth or sand".

While the number of public goods has increased over the centuries, the confusion between everybody and nobody has been perpetuated. As remarked by Grotius (1608), "in the legal phraseology of the Law of Nations, the sea is called indifferently the property of no one (*res nullius*), or a common possession (*res communis*), or public property (*res publica*)". Judith Fitzpatrick (1991) writes that,

V. Pop, *Who Owns the Moon?*, Space Regulations Library Series 4,
DOI 10.1007/978-1-4020-9135-3_5, © Springer Science+Business Media B.V. 2009

under Australian law, "fish are common property" and that "fisheries resources are publicly owned, being at once everybody's and nobody's", while Herman Haeruman (1995) speaks about natural resources that have always been "open-access resources (everybody's and nobody's property)". Andrzej Kaczmarczyk (2003, p. 3) considers that the Internet has neither a proprietor nor a ruler, belonging to "everybody and nobody as the society itself", while Ezio Manzini (2002) defines common goods as "goods' that belong to everybody and nobody in particular" and that, as long as they remain common, "cannot be reduced to marketable products and cannot therefore be bought or sold".

The category of goods above is known under different names: open-access resources, common goods, public property, public spaces, the commons, etc. Together with the oceans and atmosphere, the extraterrestrial realms are said to form part of the "global commons" – a misleading classification because these lie, physically, outside the globe. As explained by Hoffstadt (1994, p. 567), there is practically a consensus among scholars that the "province of all mankind" concept valid for the extraterrestrial realms draws upon the Roman property notion of *res communis* – i.e., "that States have rights of free access and use – but not rights of ownership – to the shared property of the community" – the same being the basis for the "freedom of the high seas". As new oceans and new shores, the outer space and the extraterrestrial realms belong to everybody and nobody or, in the convoluted Latinism of space lawyers, they are "*res communis rerum publicarum (omnium)*" (Wassenbergh, 1991, p. 61) or "*res communis omnium civicum*" (Magno, 1972, p. 165).

The *res communis/res publica*/open access regime pertains to many legal traditions. For instance, the 1839 Constitution of the Cherokee Nation provides in Article I.2 that:

> [t]he lands of the Cherokee Nation shall remain common property; but the improvements made thereon, and in the possession of the citizens of the Nation, are the exclusive and indefeasible property of the citizens respectively who made, or may rightfully be in possession of them.

The same common/public property regime exists in an anarchist society. As explained by the A-Infos Project – a group of "revolutionary class struggle social anarchists, anarcho-communists, libertarian communists, syndicalists and others who hold similar opinions", in an anarchist society "everybody and nobody" owns property:

> Land, houses, offices, machinery, factories and infrastructure are all held in common. These commodities are neither bought or sold, they are used by whoever needs to use them. . . . [I]f you don't use particular goods and property, those goods or property will become available to somebody who needs them (A-Infos News Service, 2002).

The commons regime is not monolitichal; the freedom of access varies from absolute to limited. On the right, it is bordered by *res nullius* – goods that belong to no-one and can be privately appropriated. As explained by L.F.E Goldie (quoted in Hoffstadt, 1994, pp. 567 and 588), the most lenient form of the commons is represented by *res communis*, whereby property is owned by the community as a whole, and every member has the right to use the property without having exclusive ownership rights. A stricter approach is the *res publica*, whereby the common property is centrally enclosed and cannot be used without permission from the community. On the left, it is bordered by the Common Heritage of Mankind regime, or the *res communis humanitatis*, whereby users have to share with the community the benefits accrued from the use of the commons.

In the extraterrestrial realms, there are several regimes based on the commons paradigm. The most lenient form of *res communis*, as applied by most actors to the celestial realms and most of outer space, contains nonetheless several regulations and is not strictly an open access regime. According to Article I of the OST, the freedom of exploration and use of the extraterrestrial realms pertains to States. Article VI of the OST requires non-governmental entities to obtain authorization from the appropriate State Party in order to carry out activities in the extraterrestrial realms, and to consent to being continually supervised by same. States bear international responsibility for national activities carried out in outer space and on the celestial bodies, whether these are performed by governmental entities or by private enterprise. The regime is therefore a hybrid of *res communis* – at international level – and *res publica* at municipal level, given the need for a nationally issued license. Last, but not least, the celestial bodies are considered by several States as being the Common Heritage of the Mankind – a regime going beyond the commons paradigm.

5.3 The Celestial Bodies as *Res Communis*

As stated above, most of the actors in space regard the extraterrestrial realms as *res communis*. As such, they are open for use by all States, with no discrimination whatsoever, on a basis of equality. Article I of the OST, where the above principle is enshrined, proclaims the exploration and use of the extraterrestrial realms as being the "province of mankind", and establishes a regime of "free access to all areas of celestial bodies". Other provisions in the OST confirm the *res communis* regime and elaborate it. Whereas several rules are established and ought to be respected by the State Parties to the OST, most of these remain at the stage of principle – as recognized by the OST title – "Treaty on Principles Governing the Activities of States in the Exploration and Use of Outer Space, Including the Moon and Other Celestial Bodies". Many open access regimes are self-regulatory, and where the text of the law is silent, custom is bound to develop.

5.3.1 Attributes of Res Communis

As explained by French jurist Pothier – quoted by the U.S. Supreme Court in *Geer v. Connecticut* (1896):

> The first of mankind had in common all those things which God had given to the human race. This community was not a positive community of interest, like that which exists

> between several persons who have the ownership of a thing in which each have their particular portion. It was ... 'a negative community,' which resulted from the fact that those things which were common to all belonged no more to one than to the others, and hence no one could prevent another from taking of these common things that portion which he judged necessary in order to subserve his wants. Whilst he was using them, others could not disturb him; but when he had ceased to use them, if they were not things which were consumed by the fact of use, the things immediately re-entered into the negative community, and another could use them ... These things are those which the jurisconsults called 'res communes'.

In a nutshell, *res communis* implies freedom of an actor to use a good; as long as one uses this good, one may not be obstructed by another, yet this latter is free to use the good as soon as the first has ceased to use it.

While private appropriation of the extraterrestrial realms may not exist on its fullness under a *res communis* regime, some of its elements are permitted. As stated elsewhere in this work, property is an embodiment of three attributes – *jus utendi* (the right to use), *jus fruendi* (the right to enjoy the fruits) and *jus abutendi* (the right to "abuse" one's own good). *Res communis* is built around *jus utendi* yet forbids *jus abutendi.* As to *jus fruendi* – i.e., to collect the "fruits" of the extraterrestrial realms – this is addressed elsewhere in this volume.

Under space law, *jus utendi* is not absolute; the right of using the Moon and other celestial bodies has to be carried out "exclusively for peaceful purposes" (Article IV OST, Article 3.1 MT). Article IX of the OST requires the State Parties to use the principle of co-operation and mutual assistance as a guide in the exploration and use of the outer space and celestial bodies. The State Parties need to conduct all their activities in the extraterrestrial realms "with due regard to the corresponding interests of all other States Parties", according to the same legal norm, this provision being reproduced in Article 2 of the MA. The Outer Space Treaty elaborates, up to a certain degree, the mechanism for accommodating the interests of the other members of the public. Thus, Article IX requires States Parties to publicize, "to the greatest extent feasible and practicable" the nature, conduct and locations of their space activities. If an activity by a State Party or one of its nationals is likely to cause "potentially harmful interference" with the space activities of other States Parties, the first State Party is required by the same tenet to "undertake appropriate international consultations" prior to carrying out such activity. In the same time, should a State Party to the OST rightfully fear that a space activity of another State Party could harmfully affect its own activities, the first State is entitled to request appropriate consultations on this subject. While the Treaty makes no mention of it, it is nonetheless understood that a "first-come, first-served" regime exists.

In the spirit of Article IX OST, Brooks (1968, p. 347) considered that, were a space resource so scarce that its use would endanger future uses or the needs of other States, "this situation may be an appropriation" and consultations would have to be held: "In such case the resource may be shared, or conflicting needs may be subject to settlement by agreement or adjudication". Five years prior to the enactment of the OST, Ernst Fasan (1962, p. 11) asked:

> If on Mars a very valuable pool of water is found and one state sets up a base, which depends completely on this water resource, has then another state or party got the right to extract this

> water (or a part of it) for purposes of its own and thus to coerce the other state to liquidate its base at once, or, far worse, causes the death of all inhabitants of the station in question?

The answer, under the light of Article IX, is certainly in negative, and such a situation would violate even the Lockean proviso, i.e., that one has the right of acquisition only if one leaves "enough and as good" for others (Locke, 1690, Section 26).

Article IX of the OST also forbids States Parties to cause harmful contamination of the extraterrestrial realms when studying and exploring them. Article 7.1 of the MA expands the above provisions, asking States Parties who explore and use the Moon and other celestial bodies, to prevent the disruption of the current balance of these extraterrestrial environments "whether by introducing adverse changes in that environment, by its harmful contamination through the introduction of extra-environmental matter or otherwise". Whereas no "abuse" is allowed literally by Article IX OST, nor are the other forms of *jus abutendi*. The third attribute of ownership embodies the "right to abuse" property – that is, to have a full dominion over it. In civil law, this is the right to dispose of one's property – to transform, destroy and alienate it – in other words, to use a good for a final end that cannot be reiterated by the same person (Allard, 2001, Section 3.2.2.2). *Res communis* lacks transferability, marketability. As explained by Robert P. Merges and Glenn H. Reynolds (1997, p. 121), rights can be separated in two broad classes, namely usufruct – "a right to continued use for a limited time" and fee – "a more permanent interest that can be traded, devised, or otherwise transferred". According to the two authors:

> For some purposes, the usufruct may prove to be valuable in the space environment, but generally we have in mind a fee interest, more specifically a right akin to the fee simple of Anglo-American law. The fee interest has the advantages of predictability ... and flexibility.

The absence of the fee – or *nuda proprietas* – is what distinguishes the *res communis* from the full property rights regime proposed by some authors and addressed elsewhere in this work. No right to trade extraterrestrial real estate currently exists.

5.3.2 Conflict of Users vs Conflict of Uses

A commons regime for the extraterrestrial realms must secure the harmonious cohabitation of exploitation activities with other uses of outer space and celestial bodies. Finding a way to accommodate conflicting interests and translating it into legal norms is a difficult task; as Karen Cramer (1997, p. 352) agrees, "[m]ining, astronomy, geology, solar power, manufacturing and landing facilities are not all compatible". This being said, the question arises: do certain freedoms of outer space have precedence over others? Should exploitation activities be sacrificed on the altar of the science, or vice versa? The existing *Corpus Juris Spatialis* already contains some clues, while some others may be given in line with the regimes existing for the Antarctic and even domestic regimes.

We have shown *supra* that Article I of the Outer Space Treaty enshrines the non-discriminatory freedom of exploration and use of the extraterrestrial realms, and establishes the unhindered access to all areas of celestial bodies. No prior permission is required by States Parties to conduct space activities, but Article XI of the OST requires them to make public their space activities. The UN Secretary General, one of the authorities who need to be informed according to the same article, is a passive register of priorities rather than an assigner of rights. From Article IX of the Outer Space Treaty we can infer the existence of a non-interference principle in the law of outer space; the existing principle as covered by the Outer Space Treaty is concerned with accommodating conflicts of interests between different users rather than between different uses. The OST does not establish any priority of uses of the celestial bodies; any use is allowed, provided it is exclusively for peaceful purposes (Article IV) and does not harmfully contaminate the celestial body (Article IX).

The answer is, therefore, that the laissez-faire regime enshrined in the Outer Space Treaty does not discriminate against actors and does not discriminate against activities. During the negotiations for the Outer Space Treaty, there were voices who called for a less permissive regime; according to Csabafi (1971, p. 4), the French delegate to the UNCOPUOS suggested the inclusion of a provision in the OST to the effect that priority be given to certain types of exploration and use. Three decades later, Cramer (1997, p. 354) called for setting aside historical and scientific reserves on the moon prior to making any other territorial assignments, which would include sites of historical and geological importance, as well as an area reserved for astronomical observatories. She also considered that other locations with special potential uses ought to be reserved, such as the equatorial longitude 33.1°E, seen as the ideal site for a mass launcher, as well as other areas which are most appropriate for power beaming to the Earth.

An interesting case-study has been advanced in 1994 by Jean Heidmann; the French astronomer introduced the space law community with the proposal for the establishment within the upcoming decades, of a dedicated crater on the lunar far side harbouring all high sensitivity radio astronomical observations devoted to the scientific investigation of the universe, such as SETI. The dark side of the Moon attracted interest as the ideal location for such activities, because sites on or in direct view of the Earth would be tainted by the vast man-made frequency interferences. Far from monopolizing all of the lunar far side for radio astronomy, as feared by those envisioning lunar exploration and exploitation, this proposal called for the use of "just one very specific neatly singled out crater", namely Saha – a 100 km diameter circular strongly walled crater by *Mare Smithii*. This choice for a specific single crater is seen as a means for simplifying the finding of solutions for a harmonious cohabitation with other activities for the exploration and exploitation of the Moon. The proponent urged that the space law aspects of this project be discussed by legal bodies, namely the means of initiating the discussion, the identification of the legal problems raised, their solution, the plan for a mutual understanding, and the frame for their elaboration. In pleading for possible legal solutions, Heidmann stressed humankind's "political and philosophical duty ... to provide for such a safe, unique and limited location from which SETI can be pursued for the future benefits of our global culture" (Heidmann, 1994, pp. 255–256). The extraterrestrial property rights aspects of this project are the uniqueness of this very specific site, the need for an exclusive use, the prior allocation of user rights for that particular site preceding actual occupation, and the preference for scientific activities in the detriment of other activities such as mineral resources exploitation.

Heidmann's call for legal opinions has been answered by several academics. For instance, A.A. Cocca (1995, pp. 270–273) sees the Saha project as an opportunity to invigorate the Common Heritage of the Mankind principle, embodying its first tangible application in its real interpretation. He finds nothing to impede its legal support (Cocca, 1995, p. 273) – quite the opposite, having to be recognized (ibid., p. 270). The legal problems – "[t]he establishment of the priority in the right of the facility by a registration procedure with the ITU" can be solved, in Cocca's opinion, "[b]y paying special consideration to the principles of the Moon Agreement, in particular ... the common heritage of Mankind" (ibid., p. 272). The legal framework for "affecting a crater in the Moon with specific aims – the benefit of humanity" is "sufficiently broad and complete to develop the proposal" (ibid., p. 271). Most important, Cocca considers the reservation of a lunar site for scientific activities and its utilization towards the common good of humankind as being a precedent for future utilization of the CHM principle "integrated in a whole of non-transferable and unrenounceable rights" (ibid., p. 270) and "providing useful information in the near future when other space activities will be undertaken on the Moon" (ibid., p. 272). As a procedure, Cocca envisages the start of a "discussion among international academic organizations, IAF, IISL, IAA, ICSU-COSPAR, IAU; afterwards ITU/Bureau of Radio Communications, Study Group for frequencies 1–5 Ghz", then reaching a mutual understanding by "publishing a common specific system proposal through the international academical institutions mentioned above, to be presented at the UN COPUOS and ITU" (ibid.).

Steven E. Doyle (quoted by Cocca, 1995, p. 272) suggests informal consultations with the ITU for international registration and recognition, having then "the facility identified in the international radio frequency mechanism", with the aim of establishing "priority of right of the far side facility". This would be doubled by submitting the Saha proposal to ICSU-COSPAR, the IAA and the IAU, all these phases being able to "create a historical precedent and provide for information in the near future when space activities will be undertaken on the Moon".

Francis Lyall believes that it is a matter of urgency in setting aside certain areas of the moon for scientific research. He considers that such a proposal is, broadly speaking, unobjectionable and considers the particular case of dedicating the Saha crater for SETI research as deserving "a special consideration, care and protection, whether we have a whole-lunar agreement, or a less formal arrangement" (Lyall, 1998, p. 136). He argues that the Moon Agreement as it stands does not offer enough protection for a dedicated site for radio astronomy, given that Article 8.3 of the said treaty "merely requires that activities of states parties to the Agreement shall not interfere [with] the activities of other states parties on the Moon. This is quite vague and imprecise." Lyall agrees that a "proper formal setting aside of a specific site, and its protection for radio astronomic purposes" is required, this having to be done "before any further steps are taken in Moon exploration" (Lyall, 1998, p. 137).

Patricia Sterns and Leslie Tennen correctly consider that Article I of the OST, which guarantees non-discriminatory freedom of exploration and use of the extraterrestrial realms and free access to all areas thereon, appears to run against the setting aside of a lunar area if this action would restrict the activities of states in its use and exploration. In their opinion, if unilaterally declared, such a restriction could even equate with national appropriation (Sterns and Tennen, 1999, p. 396). In the event such a zone is dedicated, they believe that other states are not prohibited by the OST from conducting activities in such areas, though they are required to seek consultations as provided by Article IX of the same document. The two authors accurately state that the rights of use and exploration are protected by the OST, but such rights appear to be linked to actual rather than projected activities. The project of dedicating a crater for radio astronomy "by unknown entities" and "at some undetermined future date" does not appear to involve the same rights as current activities, hence the OST as it stands now does not seem to directly support such an endeavour as Heidmann's proposal (Sterns and Tennen, 1999, p. 397). The two authors believe that the SAHA proposal could nonetheless benefit from another legal text, namely the Moon Agreement. Indeed, its Article 7.3 requires its States Parties to inform the UN Secretary General and other States Parties about lunar areas having "special scientific interest" so that:

> without prejudice to the rights of other States Parties, consideration may be given to the designation of such areas as international scientific preserves for which special protective arrangements are to be agreed upon in consultation with the competent bodies of the United Nations.

Sterns and Tennen (1999, p. 398) believe the above provision as allowing the protection of an area of a unique scientific importance, such as the Saha crater, even prior to a specific mission to that area.

Indeed, whereas the Outer Space Treaty does not favour any particular use, the weakly ratified Moon Agreement seems to offer priority to scientific endeavours rather than mineral resources exploitation. This is not a unique occurrence; on Earth, water use is often subject to preferences of one type of use over the other. In Kuwait, for example, the law ranks human consumption of water as having a higher priority of use than municipal and recreational use, this being more important than water use in forestation. The next uses, in order of preference, are industrial, agricultural, and private gardening (Geon, 1997, p. 293). In 1922, seven US States from the Colorado River basin entered into a compact establishing not only the "equitable division and apportionment of the use of the waters of the Colorado River System", but also the "relative importance of different beneficial uses of water" (Article I, Colorado River Compact). Thus, the priorities of use are established in Article IV of the Compact as follows:

> (a) Inasmuch as the Colorado River has ceased to be navigable for commerce and the reservation of its waters for navigation would seriously limit the development of its Basin, the use of its waters for purposes of navigation shall be subservient to the uses of such waters for domestic, agricultural, and power purposes [. . .].

(b) Subject to the provisions of this compact, water of the Colorado River System may be impounded and used for the generation of electrical power, but such impounding and use shall be subservient to the use and consumption of such water for agricultural and domestic purposes and shall not interfere with or prevent use for such dominant purposes.

In 1931, H.A. Smith (quoted by Whiteman, 1964, p. 919) offered that, in practice, it is impossible to set universal rules as to the priority of uses upon all rivers:

> In many cases it will be found that navigation interests have acquired a clear priority, both in time and in order of material importance, which cannot be disturbed except by general consent, but these instances, however numerous, do not warrant the assertion of a general rule. Even where navigation interests are prior in point of time, they may prove so unimportant that they cannot be allowed to interpose an absolute veto upon the full economic development of the river and its tributaries.

It can be seen from the above that priorities in water use are not universally set – rather, the law takes into consideration the local conditions. The lesson to be learnt in the further development of space law is that, on the extraterrestrial realms, no *a priori* preferences of use should be established, if such preferences need to be set – but, instead, the local conditions ought to be central to making such decisions.

5.3.3 De Jure and De Facto Appropriation

According to Dani Rodrik (quoted in Grossman, 2004, p. 2):

> The key word is 'control' rather than 'ownership'. Formal property rights do not count for much if they do not confer control rights. By the same token, sufficiently strong control rights may do the trick even in the absence of formal property rights.

Do control rights "do the trick" on the Moon? Are extensive user rights the equivalent of a *de facto* ownership? John R. Tamm (1968, p. 160) believes that the Outer Space Treaty did not address the matter of "what constitutes use and what constitutes appropriation". Ernst Fasan (1966, p. 54) warns that one has to draw a very clear line between the two terms – when appropriation of a celestial body or area is prohibited and the use of such celestial body or area is allowed – given that "the main effect of property is that very (exclusive) use". Gladys E. Wiles (1998, p. 523) considers that, under the current *corpus juris spatialis*, States will in fact acquire property rights in the lunar sections they actually use, "because other states will recognize that interest". She exemplifies with a hypothetical US station on the surface of the Moon, and with a Russian government desiring access to the very same portion the US station is located on:

> Practically speaking, however, Russia will not interfere with the interest of the US in that portion of the Moon for fear that the US would then retaliate in the same manner. This idea was reflected in *Johnson v M'Intosh*, where the European nations recognized the acquisitions of other nations in order to insure that their own acquisitions were recognized. Furthermore ... Russia would not take action against the US because Russia would not want the US to take action against it.

Stephen Gorove (1974, p. 30) considers that the prohibition of ownership over parts of the Moon is unfortunately a mere smokescreen, since, apart from a limited right of visitation, it allows "the very thing that ownership implies, including indefinite control and exclusion of others from the occupied area". To Wiles (1998, p. 523), the fact that both the MA and OST have articles allowing states to maintain jurisdiction and control over objects in space and on the celestial bodies makes the prohibition of national appropriation illusory.

While it is true that under the *res communis* regime no actor can usurp a prior user of a good and, in practice, possession can equate with property in this extent, this is not valid *ad infinitum. Res Communis* is the scene of a game of "musical chairs". In its Article II, the Outer Space Treaty outlaws the national appropriation of the extraterrestrial realms by any means, including "by means of use or occupation". In other words, Article II OST outlaws the application of the acquisitive prescription or adverse possession in outer space, both in its public and private incarnations. The above norm echoes the Antarctic Treaty, whose Article IV(2) states:

> No acts or activities taking place while the present Treaty is in force shall constitute a basis for asserting, supporting or denying a claim to territorial sovereignty in Antarctica or create any rights of sovereignty in Antarctica.

As stated *supra*, the exercise of authority that is both continuous and peaceful can lead, in international law, to the establishment of a historic title. The first requirement is fulfilled through the exercise of effective possession *a titre de souverain* over the territory in cause, while the second requirement is met by the lack of opposition to such display of authority (Blum, 1965, pp. 99–100). Whilst allowing the use itself, the Outer Space Treaty forbids the call upon such exercise of continuous and peaceful authority over a part of the extraterrestrial realms. Article 11.3 of the Moon Agreement elaborates:

> The placement of personnel, space vehicles, equipment, facilities, stations and installations on or below the surface of the moon, including structures connected with its surface or subsurface, shall not create a right of ownership over the surface or the subsurface of the moon or any areas thereof.

This provision strikingly resembles the Roman law regime of structures erected on the seashore. As shown *supra*, the Roman seashore was very similar in its public character with the celestial bodies; as a consequence of its use being "public by the law of nations", it was "open to anyone to put a house there to which to repair for the drying of nets or to draw up from the sea" (Justinian Institutes II.I.5.), unless it was to the injury of others. A shed, house of refuge or construction so erected was the object of private ownership as long as it stood; but once the structure fell, was removed or came down for whatever reason, the shore was wholly free again, the ground reverting to the public (Thomas, 1975, p. 75; Roby, 1902, p. 410). Hugo Grotius (1608) considers that "public things, that is, things which are the property of a nation, cannot be acquired by mere efflux of time", citing in his support Papinian:

> Prescription based upon long possession is not usually granted for the acquisition of places which are public by the Law of Nations. An instance of this is, where anyone abandons a building which he had constructed upon the seashore, or it was demolished, and another

person, having built a house in the same place, the former opposes him by an exception based upon previous occupancy; or where anyone, for the reason that he alone has been accustomed to fish for years in a certain part of a river, under the same prescriptive right forbids another to do so (Digest XLI, 3, 45).

5.3.4 Property Status of Planetary Structures

The scientific exploration of celestial bodies, as well as their industrial development, requires the placement thereon of several facilities, in the form of manned and unmanned stations and bases, various installations, equipments, instruments, and vehicles. Such artefacts – hereinafter referred to as "planetary structures" – exist in other international environments – namely in the Antarctic and on the high seas. The way the planetary structures interact with the planetary surface on whom or below which they are placed and, in some cases, with whom they are connected, raise important legal issues.

The Outer Space Treaty allows "[t]he use of any equipment or facility necessary for peaceful exploration of the moon and other celestial bodies", while forbidding "[t]he establishment of military bases, installations and fortifications ... on celestial bodies" (Article IV). Rene Mankiewicz (1968, pp. 163–164) wrote that, whereas the freedom of exploration use of the extraterrestrial realms implies the right to establish stations for these purposes, this does not result in an exclusive use of such stations by the State having established them. In his view, the establishment of a permanent station on the Moon for the exclusive use of one State or a public or private entity would contravene the OST as it "implies and in fact constitutes occupation and appropriation of the land on which it is built". Eugene Brooks (1969, p. 169) too considered that the building of "extensive permanent facilities" on the Moon, such as the "a chain of hotels and supporting facilities" would conflict with the non-appropriation principle of Article II OST as they would "take up an inordinate amount of terrain" and with the "benefit of all countries" principle of Article I as they would only serve "private individuals of one nation or a small set of nations".

The Moon Agreement left to rest the above questions by expressly entitling State Parties to establish "manned and unmanned stations on the moon" (Article 9.1) and further regulating such establishment. Thus, a State Party establishing a station is required by the Moon Agreement to "use only that area which is required for the needs of the station" (Article 9.1) and to install the station "in such a manner that they do not impede the free access to all areas of the moon by personnel, vehicles and equipment of other States Parties conducting activities on the moon" (Article 9.2). Whereas no permission is required for the establishment of a lunar station, Article 9.1 compels State Parties to promptly inform the UN Secretary-General of the location and purposes of that station upon establishment.

Given that land is everlasting while a building is not, Roman law yielded the principle *superficies solo cedit* – "the building belongs to the land" (Simpson, 1976, p. 261) or "a building becomes part of the ground". This means that the buildings erected on foundations placed in the soil become themselves, legally, part of that land (Gray, 1993, p. 7). This principle has not been carried on in space law. Instead, Article VIII of the OST provides that:

> [o]wnership of objects launched into outer space, including objects landed or constructed on a celestial body, and of their component parts, is not affected by their presence in outer space or on a celestial body or by their return to the Earth.

Article 12.1 of the Moon Agreement reiterates the above norm, providing that:

> [t]he ownership of space vehicles, equipment, facilities, stations and installations shall not be affected by their presence on the moon.

An important question brought forward by the academia is whether a moveable asset set up on a celestial body becomes a fixture, conferring thus to its owner a title to land and to its State a "right of sovereignty by occupation over the celestial body" (Nauges, 1979, p. 270). In the same line, John R. Tamm (1968, p. 160) asks whether an abandoned planetary station is free for use and occupation by other states or nationals and whether it may be destroyed or removed, considering that the OST did not address this issue. The answer to this question has been answered in the section *supra.*

As stated elsewhere in this volume, property boils down to control over access, in other words, to the right of excluding others from using a good. Sir Edward Coke (1628) stated that "A man's house is his castle – *et domus sua cuique est tutissimum refugium*". To what extent an astronaut's station is his castle – or, to what degree can an entity control the access of others to a planetary structure belonging to it? The Outer Space Treaty, in Article XII, provides that:

> [a]ll stations, installations, equipment and space vehicles on the moon and other celestial bodies shall be open to representatives of other States Parties to the Treaty on a basis of reciprocity.

The article is not clear as to how this reciprocity clause functions, nor does it state the nature of the visit. It does however require such representatives to give "reasonable advance notice of a projected visit", so that "appropriate consultations may be held and that maximum precautions may be taken to assure safety and to avoid interference with normal operations in the facility to be visited".

The visitation clause has its roots in the 1959 Antarctic Treaty, whose Article VII (3) opens "[a]ll areas of Antarctica, including all stations, installations and equipment within those areas, and all ships and aircraft at points of discharging or embarking cargoes or personnel in Antarctica" to inspection, at all times, by designated observers.

Eugene Brooks (1969, p. 169) lauds the provisions of the Outer Space Treaty, considering on site inspections of planetary structures highly desirable as means of assuring compliance with the non-appropriation principle of Article II, with the prohibition of military uses and that of harmful contamination. He also considers Article XII a warrant assuring that private activities have a proper governmental authorization and supervision, and that no interference with activities of other States occurs. Frans von der Dunk (2007, p. 253), in contrast, considers that "in principle an obligatory openness to prying eyes results", to this extent "full and uninhibited enjoyment of the ownership of hardware, software and knowhow, whether by public or by private entities" being also at issue, although the condition of reciprocity mitigates the impact of the visitation clause of Article XII.

The Moon Agreement (Article 15.1) adds to the list of open structures listed in the OST "all ... facilities" and omits the clause of reciprocity, while preserving the "reasonable advance notice" by "[s]uch States Parties" clause and its contents. The stated aim of this provision is to enable a State Party to verify that the activities of other State Parties on the Moon are compatible with the norms of the Agreement.

Scott Pace (1998) laments that Article 15 of the Moon Agreement turns the celestial bodies into "regions of lesser liberty than parts of the earth" – given its lack of respect for privacy and the "rights of persons to be secure in their dwellings" – liberties enjoyed in the democratic world. In his view, the reach of this particular norm has no limits, the inspection rights presumably extending to private quarters and personal effects and documents. Whereas such a regime, based on reciprocity, would be more of a simple requirement in the event of state-owned facilities and state employees, Pace considers that a "broad inspection requirement" like Article 15 of the Agreement would supersede the privacy rights should facilities be privately owned and occupied by private citizen – a situation, considers he, permitted by the Moon Agreement. Stephen E. Doyle (1998, p. 124) sees Article 15 of the Agreement as a species of a "general warrant" permitting anyone who reaches the Moon to "gain access to whatever it is that anyone else on the Moon is doing there" and creating thus "a people's police force by permitting unlimited inspection after the giving of reasonable notice of a planned visit". Yet David S. Myers (1980, p. 52) sees a loose interpretation of Article 15.1 of the Moon Agreement as allowing States to conceal activities "incompatible with the provisions of the agreement, especially those which relate to natural resources".

Whereas the "reasonable advance notice" by "[s]uch States Parties" clause and its contents is preserved in Article 15.1 of the Moon Agreement, the 1979 document provides for two instances where this does not apply. One such case in point concerns "persons in distress on the moon", to whom States Parties are required to "offer shelter in their stations, installations, vehicles and other facilities" by virtue of Article 10.2. The other instance is "an emergency involving a threat to human life", in this event a State Party having the outright permission, by virtue of Article 12.3 of the Agreement, to "use the equipment, vehicles, installations, facilities or supplies of other States Parties on the moon". Such use needs to be promptly notified to the UN General Secretary or the State Party concerned, according to the same legal norm. This is a considerate provision, which puts human life above property. Ryszard Hara (1983, p. 166) notes that the above norm does not address the issue of compensations in the event of misuse or overuse of its provisions; he therefore advocates the creation of a UN-sponsored expert body of international lawyers, medics and space researchers that would serve as an advisory board to States in event of disputes pertaining to the said article. Stephen E. Doyle (1998, p. 123) views the lack of "constraints of reasonableness and prior notice" in the above provision as having "many implications which appear not to have been thought through very well", having "potential for possibly life threatening interference".

In order to avoid both abuse and criticism of the above provisions, the practice of British "bothies" and Finnish "autiotupa" – or "wilderness huts" – could provide a solution for astronauts in distress. These open dwelling shelters for temporary accommodation can be found along backpacking routes and in wilderness areas, and are available for anyone to use at no charge. Although rent-free, use of such facilities is subject to an unwritten etiquette, such as using the hut's food and emergency supplies only in really urgent situations, replacement of used supplies, and greater rights to use the cottage by the last person to arrive than those already there. (Wikipedia contributors, 2006) It is hoped that purpose-built "lunar huts", or unsophisticated shelters with survival kits aimed at astronauts in distress, will become a part of the lunar landscape when practicable.

5.4 *Res Publica* and the Public Trust Doctrine

In 1997, the United Nations Secretary-General Kofi Annan declared in his Report to the General Assembly that the historic mission of the United Nations is "not merely to act upon, but also to expand the elements of common ground that exist among nations" and, "over time to convey to future generations the material and cultural heritage that we hold in trust for them" (Annan, 1997, para. 1). Admitting that the UN was created mainly with the aim of serving Member States, Annan deemed that the Organization expresses also the highest aspirations of humans around the world to achieve a just world order. In this sense, he advanced "a new concept of trusteeship", advocating the reconstitution of the UN Trusteeship Council as "the forum through which Member States exercise their collective trusteeship for the integrity of the global environment and common areas such as the oceans, atmosphere and outer space" (Annan, 1997, paras. 84–85).

The idea of a trusteeship over the commons is relatively new on the plan of international law, but has been known for a long time in several municipal law systems as the "Public Trust Doctrine". Through the collective trusteeship concept, the United Nations Secretary-General sanctioned the municipal history of the Public Trust Doctrine and extrapolated it in the international context.

5.4.1 The Municipal Dimension of the Public Trust Doctrine

The Public Trust Doctrine holds that title to certain estates is held by the State, as sovereign, in trust for the people The assets subject to public trust are most of the times the commons and essential resources, and it is the purpose of the Public Trust to "preserve resources in a manner that makes them available to the public for certain public uses" (Bray, 1999). The Public Trust Doctrine is consequently a management model for the commons, whereby the State takes an active role in regulating the access to public goods rather than relying on the good faith of the users. In this aspect, it represents a shift from the *res communis*.

James P. Power (1995, pp. 419–420) defines the Public Trust Doctrine as a legal concept representing the interest or rights of the public in natural resources; he remarks nevertheless the lack of agreement regarding the nature, purview, applications or immutability of that interest. Indeed, as Paul M. Bray (1999) notes, the Public Trust paradigm is an "historical and currently evolving concept relating to the ownership, protection and use of essential natural and cultural resources".

The institution of "trust" is best known on its private law dimension, where it is viewed as "the legal relationship which arises when estate is owned by two persons at the same time, the one being under an obligation to use his ownership for the benefit of the other" (Mackenzie Stuart, 1932, p. 1). This relationship is constituted by the truster; the trustee is under the obligation to use his ownership for the benefit of the beneficiary. Both the trustee and the beneficiary are owners of the estate, but while the legal title to the estate is vested in the trustee who has trust ownership, the interest vested in a trustee is purely formal; the beneficial ownership belongs to the beneficiary (Mackenzie Stuart, 1932, p. 1).

On its public law dimension, the equitable title to trust lands vests in the public at large, while the legal title vests in the State, restricted only by trust (cf. *Milwaukee v. State*, 1927, at 830). The Public Trust Doctrine is rooted in the Roman law, where the Institutes of Justinian (Book II, Tit. I, 1) allowed access to the seashore to everybody on the basis that the air, the running water, the sea and the shores of the sea are, by the law of the nature, common to all. According to Power (1995, p. 419), Roman law was distinguishing between navigation and fishing rights, on a hand, from other governmental property on the other hand, in so far as the first were preserved for the public and could not be diverted to private interests. This idea has been drawn upon by the Anglo-American common law system; in England, while the sovereign "maintained the titles to land, the public's rights restricted the sovereign's capacity to divest or even use the land" (Power, 1995, p. 419). Furthermore, courts in the United States have applied the doctrine (Bray, 1999).

In its American dimension, there is no general agreement regarding the precise extent of the resources hold in public trust for the people of the United States. Ronald W. Tank (1983, p. 71) maintains that the national natural resources of the United States "are to be held in trust for the full benefit, use, and enjoyment of the citizens of the United States". He describes the Public Trust Doctrine as embodying the judicially enforceable right of a citizen to assert an interest in the public lands of the United States as common property held in trust for use by all people, not only of this generation but of future generations as well. Greg Brown (1999) explains that the Public Trust Doctrine:

> provides that public trust lands, waters, and living resources in a State are held by the State in trust for the benefit of all the people, and establishes the right of the public to fully enjoy public trust lands, waters, and living resources for a wide variety of recognized public uses.

According to Bray (1999):

> title to lands underlying tidal and/or navigable waters are held by the State in its sovereign capacity as trustee for the benefit of the citizens of the State who have the right to use the waters and adjacent land for navigation and to "fish, hunt, or bathe....

In the same time, the Public Trust Doctrine is incorporated in some State Constitutions, as it is the case with Hawaii, where "[a]ll public natural resources are held in trust by the State for the benefit of the People" (Bray, 1999).

In its British form, the Public Trust Doctrine is mirrored in the institution of Crown ownership, where rights to the *regalia majora* are held by the Crown in trust for public purposes and are inalienable. Thus, for instance, the sea below the foreshore and within the territorial limit is the property of the Crown in trust for the public rights of fishing and navigation, and the rivers which are navigable and tidal belong to the Crown in trust for the public's right of passage. The British Crown also possesses rights to *regalia minora*, but in this case, the property rights possessed by the Crown can be alienated to others (Robson And Miller, 1991, pp. 70–71).

It is generally agreed that title to public trust land is vested with two co-existing titles or interests:

- *Jus Publicum*, or the equitable title, vests in the public at large, and is constituted of the main bundle of trust rights belonging to the public to use and enjoy trust lands for such public purposes as commerce, navigation, fishing and related purposes. *Jus publicum* interests are inalienable, the State being unable to convey them into private ownership or to abdicate trust responsibilities.
- *Jus Privatum*, or the legal title, vests in the State, restricted only by trust, and is constituted of the subservient proprietary rights that may exist in the use and possession of trust lands. This may be conveyed by the State to private owners, but this private interest is deemed to be subservient to the *jus publicum* (Brown, 1999; Bray, 1999).

5.4.2 The Extraterrestrial Dimension of the Public Trust Doctrine

The application of a municipal common law institution to the international arena may be met with reluctance by some critics; nevertheless, as Juma and Ojwang (quoted in Bray,1999) remark, the Public Trust Doctrine is not a peculiarity of modern law, having governed the management of natural resources – especially land – for generations; this doctrine is in fact central to traditional juridical systems, being not alien even in African society.

Although initially the Public Trust Doctrine in its Anglo-American dimension applied mainly to waterways and protected the rights of the public to use them for navigation and to fish, hunt, or bathe, the Doctrine has been since extended to protect other public values and other natural resources as subject to "intergenerational equity" (Bray, 1999). As Power (1995, p. 420) agrees, the Public Trust Doctrine is not to be considered fixed or static, but should instead be "moulded and extended to meet changing conditions and needs of the public it was created to benefit". In this particular case, the changing conditions in the exploration of outer space and the needs of the international public may call for adding an extraterrestrial dimension to the Public Trust Doctrine.

According to Mackenzie Stuart (1932, p. 5), in the creation of a trust "the use of the word 'trust' is not necessary", as "[a]ny words from which the intention can reasonably be inferred will suffice, such as 'for the benefit of' ... or other similar expressions". The Outer Space Treaty does use such vocabulary when consecrating the exploration and use of the extraterrestrial realms "for the benefit and in the interest of all countries" (Article I). The Outer Space Treaty identifies the *corpus* of the trust, but fails to identify the trustee.

The legal title – or the trust ownership – is vested in the trustee, which, in the case of the Public Trust, is the sovereign. On its domestic dimension, this does not raise many practical questions; in the case of the extraterrestrial realms, the obvious question is: "Who is the sovereign in outer space"?

As outlined in the previous chapter, the non-appropriation principle of the Outer Space Treaty outlaws national appropriation of the extraterrestrial realms; no single State can act as a public trustee for the sole benefit of its nationals. George Dietrich and William Goldstein consider that, nevertheless "there will never be a satisfactory 'mechanism' invented by human beings other than a sovereign body itself", and propose therefore "the assertion of United Nations 'sovereignty' over those areas off limits to national appropriation" as the only resolution to the "paradoxical set of principles" of the Outer Space Treaty. The UN is seen by the two authors as "the logical and natural [sovereign] body", and the collective trusteeship envisaged by Kofi Annan is considered an "international sovereignty" of the world over outer space, or a "pseudo-sovereignty" (Dietrich and Goldstein, pp. 9–11).

The equitable title – or the beneficial ownership – is vested in the beneficiary, which, in the case of the public trust, is the public at large. "The public" is a very generic category and it lacks an appropriate legal definition. In the case of the public trust over the extraterrestrial realms, the identification of a precise beneficiary seems confuse at best, contradictory at worst. While Article I of the Outer Space Treaty consecrates the exploration and use of the celestial realms "for the benefit ... of all countries", the preamble of the same document expresses the belief of the States Parties to this treaty that the exploration and use of outer space should be carried on "for the benefit of all peoples". Furthermore, while the Annan (1997, para. 1) doctrine states that the material and cultural heritage is hold in trust for "future generations", Article 11.1 of the Moon Agreement classifies the Moon and its natural resources as the "common heritage of mankind". The enumerated categories are co-fused in the very generous notion of "the public at large".

Another facet of the public trust, whose division occurs on a different dimensional plane, is hinted by the classical form of the trust institution, whereby any person may be a beneficiary (Mackenzie Stuart, 1932, p. 66), be it "capax or incapax, . . . a pupil, an insane person, a child *in utero* – i.e., an unborn child. Humankind is viewed by Dupuy (1983, p. 484) not only in its spatial dimension (international community) but also in its historical one, being "tomorrow even more than it is today". The unborn section of the mankind is, in this vision, the beneficiary of the trust, whereas the living are the trustees – very much in line with the position consecrated by the US National Environmental Protection Act which recognizes that each generation is the trustee of the environment for succeeding generations (Bray, 1999). All generations share the global commons; as stated by Edith Brown Weiss (quoted in Frankman, 1993):

> [E]ach generation receives a natural and cultural legacy in trust from previous generations and holds it in trust for future generations. This relationship imposes upon each generation certain planetary obligations to conserve the natural and cultural resource base for future generations and also gives each generation certain planetary rights as beneficiaries of the trust to benefit from the legacy of their ancestors.

Myron J. Frankman (1993) considers that the above rendering of the Common Heritage principle in terms of intergenerational justice – different from the one commonly accepted by the developing countries – may offer greater promise. Indeed, rather than a formal institution, the intergenerational public trust is more of a symbolic obligation pertaining to the living – be them from the developed or the developing world; as a non-institutional stewardship, it is the duty of every individual to carry on this mission as best dictated by the interest of the future generations.

The Public Trust has been found by the American courts to be "both active and administrative", as "it requires the lawmaking body to act in all cases where action is necessary, not only to preserve the trust, but to promote it" (Milwaukee v. State, at 830). In other words, the duty to conserve and protect the trust estate is balanced by "the duty to invest", the trustee having the power to sell and to grant leases (Mackenzie Stuart, 1932, pp. 210, 231 and 260)·The Public Trustee is therefore under an active duty to maximise the benefits that the public could take from the enjoyment of the trust corpus. In S. Pal Asija's words, "[a]s trustees of the planet Earth, we owe it to our posterity not only to maintain but to enhance the trust" (First Millennial Foundation, 1997). Indeed, for the sake of the next generations, the cosmic *res publica* should be settled, offering the unborn section of the humankind a new home and new resources.

The active character of the public trust doctrine has a Biblical support in the Parable of the Talents (Matthew 25, 14–30), where a lord entrusts his servants with his goods according to their abilities. By his homecoming, two of the servants have doubled the talents through investment, unlike the third one who buried it and returned it in the same state. After praising the two enterprising servants, the lord reprimands the third one for not having "put my money to the exchangers, and then at my coming I should have received mine own with usury" (Matthew 25, 27).

Locke agrees with the human duty to expand that what God has entrusted upon them: God, in his view, "gave the world to men in common; but since he gave it them for their benefit, and the greatest conveniences of life they were capable to draw from it, it cannot be supposed he meant it should always remain common and uncultivated" (Locke, 1690, Section 34); God has given humans "reason to make use of it to the best advantage of life and convenience. The earth and all that is therein is given to men for the support and comfort of their being" (Locke, 1690, Section 26).

The earth is not the only corpus of the trust; its meaning expands beyond its boundaries, into the Universe. In 1956, Pope Pius XII, addressing at Castel Gandolfo to the delegates of the 7th International Astronautical Congress, said:

> The Lord God ... did not place any limit on ... [man's] efforts at conquest when He said 'Subdue the Earth' (Gen. 1, 28). It was rather the whole of creation which He offered for the human spirit to penetrate... (Cheng, 1957, p. 505).

The active character of the public trust over the extraterrestrial realms means they are not to be left pristine in their entirety, the trustees being under the obligation to make the best out of the extraterrestrial realms, even if this means their exploitation and privatization. It is not a sin for humankind to expand in the universe and make use of it to the best advantage of life – to exploit the extraterrestrial resources and terraform other worlds; it is instead a divine command to put the talents entrusted upon humankind to good use. Thus, the legal regime for the extraterrestrial realms needs to be the most appropriate for maximizing the benefits of the present and future generations, and in our opinion, the active character of the public trust needs to prevail upon the other characters. An "alienation" of the public trust corpus does not run against its aims, as long as it benefits the public for whom the trus is meant.

As it will be shown *infra*, much of the public land of the United States has been opened for homesteading in the past. While in 1976 a paradigm shift occurred towards keeping the public land, as a rule, in public ownership, some land can still be sold. According to the us US Bureau of Land Management (2000), public lands can be sold if they meet one of three criteria:

1) they are scattered, isolated tracts, difficult or uneconomic to manage;
2) they were acquired for a specific purpose and are no longer needed for that purpose; or
3) disposal of the land will serve important public objectives, such as community expansion and economic development.

A homesteading of the celestial bodies – as detailed *infra*- would be in accordance with the public trust as it would contribute to the advancement of the human society through their development, while "preservation" of the extraterrestrial realms against use would run against the public trust.

5.4.3 The United Nations and the Public Trust

As stated *supra*, in 1997 the UN Secretary General advocated the reconstitution of the UN Trusteeship Council as "the forum through which Member States exercise their collective trusteeship for the integrity of the global environment and common areas such as the oceans, atmosphere and outer space" (Annan, 1997, para. 85). Beyond suggesting a *lex ferenda* role for the Trusteeship Council, Kofi Annan effectively proclaimed that the United Nations holds, *de lege lata*, a material heritage in trust for the future generations (Annan, 1997, para. 1).

Annan's call for transforming the Trusteeship Council in the forum through which States exercise their collective trusteeship for the extraterrestrial realms as part of the commons is the culmination of forty years of proposals coming from both the academic circles and civil society. As early as 1956, Walter Nash (quoted in Jenks, 1958, p. 367) – a leading New Zealand politician – proposed that Antarctica should be placed under the trusteeship of the United Nations. Two years later, Jenks (1958, p. 397) considered as "reasonable" that nationally appropriated extraterrestrial territories "should be subject to the international trusteeship system". Not much later, Cyril Horsford (quoted in Woetzel, 1962) recommended the placement of the celestial bodies under an international trusteeship system set up by the UN and modelled on the existing Trusteeship Council. This new specialised agency, established under Article 59 of the UN Charter, would administer the celestial bodies through states, as under the present trusteeship on earth.

The Trusteeship Council itself was created by the UN Charter, whose Article 75 provided for the establishment under the authority of the United Nations an international trusteeship system for the administration and supervision of trust territories. Article 76 shows that the focus of the trusteeship system was the population of such territories and not the land itself. The next article lists the category of territories where the trusteeship system shall apply, areas as may be placed thereunder by means of subsequent trusteeship agreements that would establish the terms of the trusteeship:

a. territories now held under mandate;
b. territories which may be detached from enemy states as a result of the Second World War; and
c. territories voluntarily placed under the system by states responsible for their administration.

It is to be noted that it was not the perception of a need for a public trustee for the commons that brought the issue to the attention of the UN General Secretary, but the redundancy of the Trusteeship Council once with the 1994 independence of Palau, the last trust territory (O'Donnell and Goldman, 1997, p. 319). In November of that year, the Report of the Commission on Global Governance (1994, Chapter 7) called for the Trusteeship Council to be given "a new mandate over the global commons in the context of concern for the security of the planet". Maurice Strong (1995) considers that:

> [i]n practical terms, the Council would exercise trusteeship over outer space, the oceans, and the various environmental treaties and agreements related to the care and preservation of the integrity of major global ecosystems.

One year later, Joseph Cassar, the Maltese representative to the UN General Assembly, submitted a draft resolution in the Sixth Committee requesting the Secretary-General to report to the 51st Session of the Assembly after having sought the views of the Member States on the future of the Council. While some voices in the UN General Assembly were calling for the disbandment of the Trusteeship Council, Cassar employed a different tone by restating the proposal made by his government in 1990, during the 45th Session of the Assembly that, "in addition to its role under the Charter, the Trusteeship Council hold in trust for humanity its common heritage and common concerns" (United Nations General Assembly, 1995). The 1995 Maltese proposal was in the sense that "consideration be given to designating the Trusteeship Council as a trustee of the common heritage of humankind to ensure the necessary coordinated approach to this matter of common concern" (Morris, Bourloyannis-Vrailas 1996, p. 500).

The Annan proposal intended to concentrate the *jus privatum* of the public trusteeship in the hands of the United Nations Trusteeship Council. While Strong (1995) deemed the mandate over the commons as seeming to be "the ideal function for the Trusteeship Council", the view adopted by George Dietrich and William Goldstein (1998, p. 9) was more reserved: – "It remains to be seen how the member states will react" to the proposal of the UN Secretary-General.

Despite having a measure of governmental and non-governmental support, the plan of entrusting the UN Trusteeship Council with the global commons did not endure, however. Less than a decade later, the same Secretary General deemed that, "[j]ust as we contemplate creating new institutions . . ., we should abolish those that are no longer needed, such as the Trusteeship Council" (Annan, 2005). It is likely that a UN trusteeship over the global commons would have been met with resistance by many of the spacefaring states – as it was the case with the Moon Agreement – as going far beyond its mandate.

Some authors, while agreeing with the trusteeship over outer space, do not view the United Nations as the most appropriate entity to manage the extraterrestrial public trust lands. Indeed, the Trusteeship Council was created as an instrument of decolonization, not as an instrument of space colonization. Allen D. Webber (1983, pp. 1447–1451) believes the UN to be an "inappropriate body for celestial body resource development", due to its slowness, institutional weakness, incapacity of enforcing any mandate and inability to respond effectively to conflicts among nations. In the same time, he sees the UN as unable to govern extraterrestrial resource development independent of terrestrial concerns, due to the wide range of activities dealt with by the UN; in his view, a UN regime would fail to install confidence in investors and will be "inconsistent with the principle that space should be developed for the benefit of the mankind, and not for the benefit of nations". As an alternative, he proposes a regime for exploitation of extraterrestrial resources that would be governed by "an autonomous panel of individuals, not dominated or controlled by any nationalistic entities".

An alternative to the UN Trusteeship Council as trustee of the extraterrestrial realms, the "United Societies in Space" (USIS) advocate the establishment of a "Regency". Declan J. O'Donnell (1999, p. 7) considers that, given its uniqueness, outer space is not yet ready for the full measure of self governance, being therefore in need of a transition stage before acquiring full UN membership. As a practical transitory solution, he co-founded together with Philip R. Harris the USIS, a Colorado non-profit corporation sponsoring the Regency of the United Societies in Space – ROUSIS (O'Donnell, Harris 1998, p. 271). ROUSIS is a space governance regime envisaged as a "trustee government"; its Constitution starts with a

"Declaration of Trusteeship", where the beneficiary is identified with "all of humanity and the Earth Nations in which they reside, consistent with the terms of the Outer Space Treaty", while Article I.7 proclaims that "ROUSIS shall represent all humankind as beneficiary". The *corpus* ("the res of this trust") is to consist of "space resources, the trustee having to "manage same for the benefit of all, asserting jurisdiction to that end" (O'Donnell, 1999). The ROUSIS is envisaged to "fill in the political void in outer space with a civilian governance entity", its principal mission being "to build an off world estate with a traditional space governance capacity within 100 years" (O'Donnell, 1999, pp. 2 and 7). ROUSIS is seen by its promoters as a transitional government for outer space that would operate on earth until the latest 2100, when it will wind up its affairs in favour of an "in-space governmental convention of settlers". "As settlers arrive in space and establish communities" – proclaims the ROUSIS Constitution – "then those settlers shall be deemed priority beneficiaries, consistent with the United Nations Charter of 1947 in respect to territories not yet ready for the full measure of self-governance" (O'Donnell, 1999, Annex, p. 1). Until then, a convention is called for August 4, 2000 in Denver to establish the ROUSIS, a number of 200–2,000 regents being elected to the Regency (O'Donnell, 1999, p. 7).

The activities of the Regency are based on the premise that ROUSIS is a government; as such, it is entrusted with extensive attributes in the field of property rights. Thus, ROUSIS is to have the authority to "grant, regulate and terminate common law usage rights in the nature of leaseholds and easements" (O'Donnell, 1999, Annex p. 2, ROUSIS Constitution Article 1 Section 5), being "the exclusive granter of such property rights to space resources in space subject to treaty provisions provided to the contrary from time to time". These rights are to be administered "for the common good consistent with a consensus space development plan by ROUSIS and the municipal authority related to the estate", and no lease or easement "shall exceed 99 years and each shall be defeasible by ROUSIS for cause, as well as be limited in time and space and character to meet the societal purpose for which it is granted".

As a *lex situs*, the ROUSIS legislature is to "provide appropriate procedures for the usage of space resources, the kinds of estate to be granted, the formula for consideration and fees, and the terms for holding same", and to "fix procedures for the addition of statutes and rules and policies. Both civil and criminal laws shall be made ..." (Article I.6). In its Executive Branch, ROUSIS is to have "special powers (...) to maintain and promote ... societal rights that are deemed to run with the territory and will become part of any successor government therein "(Article II.10); thus:

> A body of Common Law estates in property shall be deemed approved, such as tenancies of 99 years or less, easements, and trust estates, as well as a Common Law of contracts, torts, and criminal laws, as extended to outer space herewith (Article II.10).

Initially – prior to promoting ROUSIS – O'Donnell and Harris considered the establishment of a space "Metanation" to serve as trustee manager of the interplanetary commons for all humankind (O'Donnell, 1996, p. 3) and was to be founded – according to O'Donnell and Harris (1988, pp. 275, 278–279) – by the spacefaring nations through the UN under a "Trusteeship Council for New Territories", where spacefaring nations would act as trustees for one hundred years. Under the auspices of the Metanation, a "Lunar Economic Development Authority" was envisaged to function, being entrusted, *inter alia*, with "leas[ing] surface rights for private or public sector lunar macroprojects and collect fees therefrom". While replacing the Moon Agreement with a new "Treaty on Jurisdiction in Outer Space", the establishment of the Metanation was considered by its initiators as a "pro-space development approach" preserving the concept of Common Heritage of Mankind (O'Donnell and Harris, 1998, pp. 275, 278–279). The Metanation was viewed by its promoters as able to solve the issue of ownership of space resources, by defining and registering benefit shared resources. Thus:

> The public part is dedicated to public improvements and the rest is leased, subject to ongoing rents and royalties for the benefit of all. A system of private property will result in outer space, albeit subject to rules of equitable estates and other restrictions. No fee titles are anticipated but the equitable estates may be adequate for commercial uses, at least among users of equal status in the venue. All titles will remain inchoate, defeasible, and infirm but, nevertheless, entirely adequate for commercial purposes, much like the long term lease as an equitable estate for private property purposes (O'Donnell, 1996, p. 5).

It was agreed in the ROUSIS constitution that ROUSIS shall not be established "so long as all space faring nations object to its establishment at the Denver Space Convention on August 4, 2000 AD" (O'Donnell, 1999, Annex p. 10: ROUSIS Constitution Article VII, Section 2).

5.5 Arguments for the Commons Paradigm

The current *res communis* regime is the *lex lata* in outer space. As such, it is accepted by most of the actor on the space scene. As recognized by Andrzej Gorbiel (1985, p. 142), "the principle of the freedom of the open sea served as a justification for the principle of the freedom of outer space". The former has been defended by Hugo Grotius (1608) on the grounds that:

> [t]hose ... who deny this law, destroy this most praiseworthy bond of human fellowship, remove the opportunities for doing mutual service, in a word do violence to Nature herself.

Patricia Sterns and Leslie Tennen (2002, p. 2) favour the status quo on the basis that there are enough examples where private entities can profitably extract resources from lands they do not own, this making fee simple ownership of extraterrestrial property irrelevant. As practical examples, they give logging rights, offshore oil platforms, and grazing leases on public lands. The two authors consider that the prohibition of claims of appropriation helps to assure that all states have the right to access to outer space and enables the uniform preservation of extraterrestrial environments (Sterns and Tennen, 2002, p. 4). In the same time, the prevention of armed conflicts into space and the maintenance of international peace and security are more useful to the world than would be the abrogation of the non-appropriation principle (Sterns and Tennen, 2002, p. 5).

Alanna Hartzok (2000) believes that "the Earth is the birthright of all people" and calls for imposing "natural resource taxes" that would function as user fees and would ensure "fair and efficient use of God's gift to all". In support of this argument, she quotes Thomas Paine:

> Men did not make the earth ... It is the value of the improvement only, and not the earth itself, that is individual property ... Every proprietor owes to the community a ground rent for the land which he holds.

As men did not make outer space either, the argument carries on into the extraterrestrial realms.

5.6 Arguments Against the Commons Paradigm

The current regime has critics from both ends of the political spectrum. Whereas authors of conservative and libertarian persuasion would like to see a privatization of the extraterrestrial realm, there are voices on the left that call for sharing what has been extracted from the common pool. Many of the arguments advanced by the two sides will be presented in the following two chapters. Besides ideological arguments, the *res publica* is criticized on practical grounds.

The main tangible argument against the paradigm has been formulated by Garret Hardin (1968). His "tragedy of the commons" thesis holds that unrestricted access to a commons results in its overexploitation due to the selfish behaviour of individuals. Lawrence D. Roberts (1990, p. 167) explains that, when a parcel is assigned for common use, each person with the ability to exploit the commons will proceed accordingly, because the individual benefit received by him is much greater than the joint loss received from his removal of resources. Whereas the benefit belongs to the user, the loss is shared among many parties. This state of fact, as explained by Roberts, prompts each individual to continue to enjoy the commons far beyond the threshold where the benefit to all parties is maximized. This natural egoism has been remarked more than two millennia ago by Aristotle (350 BC, Section 2.3):

> For that which is common to the greatest number has the least care bestowed upon it. Every one thinks chiefly of his own, hardly at all of the common interest; and only when he is himself concerned as an individual.

Roberts (1990, pp. 167–168) clarifies that the commons principle is applied chiefly to fishing in the high seas, and to lands for grazing, foresting and farming – renewable resources with a limited ability to replenish themselves. The tragedy of the commons, according to him, occurs when exploitation goes beyond the renewability rate.

Quantity-wise, most of the resources of the extraterrestrial realms are inexhaustible in practice. Nonetheless, the tragedy of the commons may occur in scarce resources.

On philosophical grounds, the enclosure of the commons is seen as natural by some authors. As explained by John Locke (1690, Section 33), "God gave the world to men in common, but since He gave it them for their benefit and the greatest conveniencies of life they were capable to draw from it, it cannot be supposed He meant it should always remain common . . .". French legal scholar Pothier – quoted by the U.S. Supreme Court in *Geer v. Connecticut* (1896) – explains:

> The first of mankind had in common all those things which God had given to the human race. . . . The human race having multiplied, men partitioned among themselves the earth and the greater part of those things which were on its surface.

Michael J. Listner (2005) considers the current doctrine of space law as "restrictive and suffocating" and calls for rethinking it for the sake of progress in space. In his view, the OST is more a hindrance than a safeguard, as the *res communis* regime effectively limits innovation and expansion in the extraterrestrial realms, being an important roadblock to the exploitation and commercialization of extraterrestrial resources in the absence of property safeguards.

5.7 Conclusion

De lege lata, the extraterrestrial realms belong to "everybody and nobody", pertaining to the legal category of *res communis*, whereby use is permitted yet title is denied. *Jus utendi* is not absolute, having to be enjoyed with due regard to the corresponding interests of other users and likely users. Whilst users cannot be usurped, use does not result in appropriation by prescription. Between planetary structures and the soil of the celestial bodies there is a layer of "legal insulation" – no physical interaction between the two can change the legal status of the other.

As with every balanced situation, the equilibrium has been challenged from both sides of the political spectrum. As seen, there are voices that would like to bring the extraterrestrial commons under the sway of the United Nations. In the same time, calls have been made for an enclosure of the extraterrestrial commons. The following two chapters will evaluate these approaches.

Chapter 6
Homesteading the Final Frontier

Space: the final frontier

Gene Roddenberry

Mine is better than ours

Benjamin Franklin

Good fences make good neighbours

Robert Frost (1915)

6.1 Introduction

On January 14th, 2004, the President of the United States George W. Bush announced a new space exploration vision for NASA, launching the space agency on a "journey to inspire, innovate and discover". After a decades-long hiatus where humans were stuck in low earth orbit, the American objectives in space undertook a radical transformation. The new vision provides for a human expedition to the Moon by the year 2020 as a prelude to future human exploration of Mars and other destinations, and for unmanned expeditions across the solar system. The President urges the "use of lunar and other space resources to support sustained human space exploration to Mars and other destinations" and the robotic exploration of "Jupiter's moons, asteroids and other bodies to search for evidence of life, to understand the history of the solar system, and to search for resources" (Aldridge, 2004, pp. 2–53).

Soon after George W. Bush's speech, a Presidential Commission was created with the aim to scrutinize and make suggestions on implementing the vision. Having sought input from within governmental circles and from the public, the nine Commissioners came to fully endorse the Presidential vision, outlining as well a need for fundamental changes in the manner in which the United States approach space exploration and manage the vision (Aldridge, 2004). One of these shifts of attitude is found in Recommendation 5–2, where the Commission proposes that:

> Congress increase the potential for commercial opportunities related to the national space exploration vision . . . by assuring appropriate property rights for those who seek to develop space resources and infrastructure (Aldridge, 2004, p. 33).

This *lex ferenda* proposal would represent an important shift from the *res communis* approach consecrated in the Outer Space Treaty. Private property rights have loon been promoted by space advocates as the most appropriate – if not the only way for developing and settling outer space. In the tradition of the American Frontier, whose settlement was encouraged by governmental plans of privatizing the public

V. Pop, *Who Owns the Moon?*, Space Regulations Library Series 4,
DOI 10.1007/978-1-4020-9135-3_6, © Springer Science+Business Media B.V. 2009

domain, the proponents of the private property of space seek a similar privatization of the extraterrestrial realms. In what follows, the concept of lunar homesteading will be analysed, together with the role of private property rights in the context of space development.

6.2 The High Frontier Thesis

In 1976, Princeton physicist Gerard K. O'Neill published "The High Frontier", a book where he sketched out a course for expanding humankind into outer space on a permanent and sustainable basis. A decade prior, Gene Roddenberry, the creator of a successful TV show, had dubbed outer space "the final frontier", while in 1965, the Gemini 5 astronauts chose a Conestoga wagon as their flight emblem, true to Ralph Waldo Emerson's (1862) guidance – "Hitch your wagon to a star".

The association between space colonization and the American Frontier is a recurrent theme in the pro-space ethos, complementing the earlier view of the American continent as a "New World". At the end of the 19th Century, historian Frederick Jackson Turner explained American development through "[t]he existence of an area of free land, its continuous recession, and the advance of American settlement westward" (Turner, 1893). True to the above "frontier thesis", space activist Gregg Maryniak (quoted in Husain, 2000) believes that people raised in lands that have been settled are driven by the idea of building new frontiers, their heritage giving them a psychological urge to continue to expand. According to historian Barbara Tuchman (1976), the frontier and private property rights are deeply connected, in her view:

> [t]he open frontier, the hardships of homesteading from scratch, the wealth of natural resources, the whole vast challenge of a continent waiting to be exploited, combined to produce a prevailing materialism and an American drive bent as much, if not more, on money, property, and power than was true of the Old World from which we had fled.

As described by Alan Marshall (1995, p. 46) – who subsequently goes to criticize the paradigm – frontierism emerges from the individualism deep-rooted in the American psyche, securing as sacrosanct the operation of laissez fare-ism:

> Frontierism involves a belief in the individual to surmount the challenges of a new situation, a new territory or a new environment and carve out an existence. Once the individual has done this they deservedly call that territory or environment their own.

In a 2003 testimony before American legislators, Rick Tumlinson (2003, p. 16), founder of the appropriately named Space Frontier Foundation, pleaded for "the right to own new land in space", asserting the crucial need to "begin putting in place the rights of those who explore and develop such new 'lands' in space to own them" in order for these to live up to their potential of great sources of wealth:

> Throughout history, it has been the ability to gain and hold land which has driven [the explorers and developers] forth, and given them the will to carve new human domains out of wilderness. Space is no different. If people are going to invest their wealth and lives in opening the frontier, they should have the right to pass what they have done down to the next

generations. When the time is right, the US should stand up and recognize that in space, the same rights to won property exist as on Earth.

The frontier is inexorably linked with the idea of individualism and private ownership, of transforming *res nullius* and *res publica* into private property. Whereas Marshall (1995, p. 52) joins the frontierist paradigm of space development to the *res nullius* approach since it "sanctions a physical appropriation of extraterrestrial materials that amounts to annexation", the development of the American frontier pertains also to a privatization of the *res publica*, whereby the central authority serves not as a mere protector of acquired property rights, but also as an assigner of property rights. As it will be seen *infra*, a number of authors call for a similarly active role for the state in providing lunar homesteads.

6.2.1 The Frontier Spirit

On December 13th, 1963, the UN General Assembly adopted unanimously the "Declaration of Legal Principles Governing the Activities of States in the Exploration and Use of Outer Space", where the extraterrestrial realms were declared as "not subject to national appropriation by claim of sovereignty, by means of use or occupation, or by any other means". This incarnation of the non-appropriation principle, limiting the upward advance of the frontier, came two hundred years – almost to the day – after the British Sovereign enacted a similar policy for the American frontier.

On October 7th, 1763, King George III issued a Royal Proclamation whereby he attempted to arrest the westward advance of the frontier. By that document, the British Crown set aside under its own dominion and protection the lands lying beyond the sources of the rivers flowing into the Atlantic. The King also forbade British subjects from settling beyond that line, as well as purchasing or taking possession of the reserved lands – a measure seen as transcending his constitutional powers (*Johnson v. M'Intosh*, 1823). Defied by many, the prohibition of "crossing the Alleghenies" was condemned even in Britain by Edmund Burke (1775), who empathized with the colonists, lamenting "this avarice of desolation, this hoarding of a royal wilderness" having occurred when "the Crown not only withheld its grants, but annihilated its soil". He warned that, in any case, the people would occupy without grants, as they already have, while the authorities "cannot station garrisons in every part of these deserts", where the settlers "would wander without a possibility of restraint; they would change their manners with the habits of their life". Burke strongly criticized what he saw an attempt "to forbid as a crime, and to suppress as an evil, the Command and Blessing of Providence, Increase and Multiply" and "to keep as a lair of wild beasts, that earth, which God, by an express Charter, has given to the children of men". Instead, he called for a return to the previous policy of inviting the people "by every kind of bounty, to fixed establishments" and of encouraging "the husbandman to look to authority for his title" and "to believe in the mysterious virtue of wax and parchment".

The attempts to restrict land sales and settlement and to limit the boundaries were useless, as explained by Turner (1893); the frontier of settlement advanced steadily, carrying with it democracy and nationalism. Most important, the frontier carried and generated individualism, the wilderness transforming complex society into "a kind of primitive organization based on the family". Without opposing it, Turner saw this tendency as "anti social", producing "antipathy to control" – the tax collector being seen as an embodiment of oppression. The historian agreed with the view that the frontier conditions are some of the factors that generated the American Revolution, where "individual liberty was sometimes confused with absence of all effective government". In this approach, the frontier is seen – without being expressly identified as such – as the fief of anarcho-capitalism.

According to Klaus Heiss (quoted by Adam, 2005), director of space advocacy group "High Frontier", the United Nations are "playing King George at the time of the American revolution thinking they can tell everyone else what to do. They can't". Beth Elliott (1998), a US space entrepreneur, has the same stance when expressing her beliefs in principles of self-government as opposed to the earlier attempts of enforcing "confiscatory law from abroad on a frontier economy". She trusts that the best law for any frontier, as proven by history, is the law developed by the pioneers themselves, given their first-hand understanding of the particular conditions of the frontier. Elliott also believes that the pioneers have a personal stake in law that provides a fair mechanism for dispute resolution and that protects the individuals, their endeavours and their property from fraud and force. She criticizes "externally imposed regulatory regimes" as having an inherent interest in turning the pioneers into "cash cows", as seen in the case of the UNCLOS. Hoping to avoid this happening in space, she is very vocal in refusing to be told by other entities:

> whether individually or in some collective grouping of the non-spacefaring, where in space we may go, what we may bring back, and whether or not we may keep the products of our hard work and creativity. No matter how 'representative' of humanity any such body might be, it would still be a group of people deciding on their own to give themselves power over other people and their activities ... We do not believe anyone can claim jurisdiction over bodies beyond Earth, much less over the people using those bodies, in absentia, whether for themselves directly or on behalf of humanity in the abstract.

Whereby at first the squatter claims were not recognized, the US Congress approved in 1841 the Preemption Act, permitting many squatters on government land to purchase up to 160 acres at a nominal price before the land was offered for sale to the public. This prompted Senator Scott of Indiana (quoted by Turner, 1893) to consider the Preemption Act "merely declaratory of the custom or common law of the settlers".

The frontier spirit can translate in a dual way in the settlement of space. The first approach is that of "crossing the Alleghenies" and establishing an anarcho-capitalist way of life on the Moon, based on individualism – as shown *supra*. As with the Preemption Act and the Spitzbergen Treaty, the possession can be later offered the protection of a sovereign entity. The second approach is that of an authority privatizing the commons – as it will be shown next.

6.2.2 Partitioning the Common Heritage

In the first part of the 19th Century, the United States greatly extended their territory by means of cessions, purchases and annexations, much of the newly acquired land becoming part of their public domain. On March 11th, 1850, United States Senator William Henry Seward delivered a speech in the US Senate, where he addressed, *inter alia*, the legal status of these lands. He came to view the new territories as a "possession, to be enjoyed either in common or by partition by the citizens of the old states", due to its having been "acquired by the valor and with the wealth of the whole nation". More important for the international lawyer of our time is the precise language he used when describing the new territory as a being:

> part of the common heritage of mankind, bestowed upon them by the Creator of the universe ... We are his stewards, and must so discharge our trust as to secure in the highest attainable degree their happiness (Seward, 1850, paras. 22–23).

In suggesting the possible "partition" of the "common heritage of mankind" by its public trustees as a means of securing the "highest degree of ... happiness", Seward was being led by the frontier spirit. According to Turner (1893), the frontier greatly influenced the disposition of the American public lands, in "sharp contrast with the European system of scientific administration". The United States of the time were the scene of a deep East-West conflict. As explained by Turner:

> Efforts to make this domain a source of revenue, and to withhold it from emigrants in order that settlement might be compact, were in vain. The jealousy and the fears of the East were powerless in the face of the demands of the frontiersmen ... [A] system of administration was not what the West demanded; it wanted land.

President John Quincy Adams admitted to have failed in his task of implementing a "system of administration, which was to make the national domain the inexhaustible fund for progressive and unceasing internal improvement". While initially a bill allocating the income from the sales of the public lands to all the States of the Union passed both houses of the US Congress, it was vetoed by President Jackson in December 1832, his recommendation being that "all public lands should be gratuitously given away to individual adventurers and to the States in which the lands are situated" (Turner, 1893).

This ethos was furthered in 1844 by George Henry Evans and John Cummerford, the promoters of the "Land Reform movement", who envisaged providing free land as a means of attracting westward the redundant population of the East. Their programme advocated granting every poor man homesteads comprising 160 acres each of the vast *ager publicum*, whilst prohibiting the disposal of the public domain in large blocks to speculators (Robbins, 1942, Chapter VI). Another proponent of the Land Reform movement was Horace Greely, who supported "man's natural right to use any portion of the earth's surface not actually in use by another", given that "nature indicates and justice requires ... equal opportunities to all". In 1850, Sam Houston of Texas contended that the actual settlement of the West had a bigger value than the amount which the government could gain from the sale of the lands, and invited the Committee on Public Lands to:

> inquire into the expediency of granting to each [poor] family . . ., citizens of the United States, or emigrants who are now here or may arrive previous to March next, 160 acres of land (Robbins, 1942, Chapter VII).

The Land Reform movement finally won with the passage, in 1862, of the Homestead Act, whereby the Government provided grants of 160 acres to settlers, who were required to cultivate and live on the land for five years in order to receive tithe to it (Wiles, 1998, p. 515). After six months of residency, homesteaders were provided with the option of purchasing the land from the government for $1.25 per acre.

The Homestead Act was administered by the General Land Office, a specialized agency of the U.S. Department of the Interior. Created in 1812, the GLO would also survey and sell the public domain in the West. In 1946, the GLO would merge with the Grazing Service to become the Bureau of Land Management, an agency of the same Department. Throughout its history, the GLO/BLM would also be approached by individuals interested in a possible extension of its jurisdiction to the Moon.

6.2.3 Homesteading Prior the Non-Appropriation Principle

In August 1946, R. L. Farnsworth, president of the United States Rocket Society Inc, called for his country to occupy the Moon (The Sheboygan Press, 1971). Two years before, he had exchanged letters with the General Land Office of the United States Interior Department, enquiring as to the competent agency and the procedure for homesteading the Moon. One month after having received the unusual query, Land Office Commissioner Fred W. Johnson replied:

> New lands which may be designated as public domain under the sovereignty of the United States are subject to the jurisdiction of the General Land Office. Consequently, those portions of the Moon which may become public domain are susceptible of acquisition under the more than 5,000 public land laws administered by this office, relating to mineral developments, homesteading, grazing or other type of land use.

There was, nonetheless, an important caveat:

> If acquisition of the lands is to be sought under the homestead law, application therefore must be accompanied by an affidavit declaring that the applicant has personally examined the land and is well acquainted with its characteristics. The applicant must be prepared to establish his permanent residence upon the land to the exclusion of a home elsewhere, within six months after his entry has been approved (Associated Press, 1944).

Two years later, in January 1946, Joel David Wolfsohn, assistant land commissioner, detailed the procedure to a journalist, explaining that a lunar homesteader would have to personally go and study a piece of lunar land, and subsequently file a claim with the nearest land office. He would then have to offer some form of evidence that he wants to live there: "The best way is to take along the wife and kiddies. A couple of cows, too, would give the undertaking an air of permanence." While Wolfsohn agreed that the GLO did not yet have a lunar division, nor do the public land laws deal with our celestial satellite, he was however confident that, if and when the need arises, the GLO and Congress "will not be caught snoozing" (Associated Press, 1946).

On October 17th, 1946, the acting Director of Land Management replied by means of a letter to a claimant having asked for a lunar homestead:

> If land on the Moon should at any time be officially proclaimed to constitute public land of the United States as a result of sovereignty obtained through exploration or other means, arrangements will doubtless be made at that time to entertain such applications as yours – until such arrangements have been officially established, however, this bureau has no authority to receive applications for land on the Moon (Statesville Landmark, 1946, p. 1).

The letter from Fred Johnson then continued:

> Should the United States lay claim to the Moon and throw it open to homesteading, you must certify that you have personally examined the land and are familiar with its characteristics before acreage can be granted (Johnson, 1946).

In 1948, Robert D. Eaton and Charles W. Honhold, two young men from Pennsylvania, approached the Interior Department with a claim for the Moon. In a formal reply dated March 31st, Secretary of Interior J.A. Krug explained that the sovereignty of the United States has never been established on the Moon, and that in any case federal homestead laws require applicants to inspect in person the area sought:

> So far as we have been advised, no laws have yet been enacted by Congress authorizing the receipt and action upon ownership claims to the Moon or other planets such as you have in mind (Associated Press, 1948).

Overwhelmed with requests from people wanting to stake claims to the Moon, the Department of the Interior had to draft a standard form letter to send as a reply to the would-be Moon settlers:

> It is not now nor has it ever been possible for anyone to make application for or to obtain the right to land on the Moon or planets through the United States government. The United States has as yet made no claim to any territory beyond the earth (Keller, 1959).

A later incarnation of the form letter sent by the BLM explained that there is no legal basis to regard the celestial bodies as American public land – hence "there is no means or method by which they may be officially claimed or obtained by anyone" (Carson, 1965).

By 1965, more than two hundred extraterrestrial claims and queries had been lodged with the BLM (Carson, 1965). After two decades of denying all lunar claims because no one had been able to establish the necessary six months of residence there, the Bureau of Land Management started to refer all inquiries to the Office of Territories in the US Department of the Interior. In May 1965, Ruth G. Van Cleve, director of the Office of Territories announced that, since her department handles areas non-contiguous to the United States, it is natural that "both law and good sense compel another jurisdictional arrangement" for the Moon (Stevens Point Daily Journal, 1965).

In 1967, with the ratification of the Outer Space Treaty, the United States formally agreed not to advance any territorial claims in outer space. This meant, in practice, that no part of the extraterrestrial realms could become its public domain, and that it could not endorse a private appropriation of the same. The Homestead Act also expired in 1976, making impossible any *lex lata* application of the frontier paradigm to outer space. Nonetheless, the advocacy for homesteading the Moon continued, as it will be further seen.

6.2.4 Coelum Clausum: Abrogation of the Non-Appropriation Principle

While a number of authors and pro-space activists believe in the possibility of property rights existing independent of sovereignty, many are convinced that the non-appropriation principle of the Outer Space Treaty is an obstacle that ought to be abrogated in order for property rights to survive – or that the OST be abandoned altogether.

Prior to the advent of the Space Age and throughout its initial stages, many authors tackled the issue of space sovereignty, a number of them being favourable to the paradigm before the zeitgeist would move from colonialism to decolonization. We have shown in the first chapter how Alexander the Great wept at his inability to conquer one – let alone several worlds, and how Cecil Rhodes decried the stars being too far to reach, hence impossible to annex. Less than six decades after Rhodes' lamentations, a human would pierce through the celestial vault and would make such conquests possible.

The ethos behind the annexationist approach was twofold. Even more than being coveted for its important riches, at the beginning of the cold war, the Moon was seen as an important military outpost, the ultimate "high ground", and scientific endeavour in outer space was under the shadow of military uses. Farnsworth's call for a US annexation of the Moon as early as 1946 was motivated by his desire that his country become "the superpower of the atomic age" (The Sheboygan Press, 1971). In January 1958, US Brig. Gen. Homer A. Boushey expressed the view that missiles could be better guided and observed if fired from a lunar outpost rather than from earth, and urged the establishment of such a missile base on the Moon, possibly on the far side. One year later, the US air force director of guided missiles, General Dwight Black, told the US Congress that the first nation who lands on the Moon will "probably have a tremendous military advantage over any potential enemy". In order to establish such advantage, the air force contemplated a "Lunar Strategic System" which might comprise a "military bombardment retaliatory capability from a moon base" (Richelson, 2000).

It is important to note that the military uses of the Moon had territorial implications. In order to turn the Moon into a military base, Gen. Boushey pointed out that the Moon would have to be US property and that his country cannot afford to come second "in a territorial race of this magnitude". Gen. Black implied as well that his country would have to annex the Moon and keep the Soviet Union from landing on it (Richelson, 2000). In March 1962, Dandridge Cole, analyst for the missile and space vehicle department of General Electric, voiced his fears that if USSR establishes bases at each Lunar pole and four bases on the Lunar Equator, "they could then lay claim to the whole moon and threaten to shoot down any aggressor" (Associated Press, 1962).

With the advent of military satellites, the strategic importance of the Moon would however diminish. As noted by Jeffrey Richelson (2000):

> the creation of a strategic triad, particularly the buildup of significant land and sea-based missile forces, along with sophisticated warning systems, served to confront Soviet strategic planners with the knowledge that an attack on the United States would result in "sure and massive retaliation," even without U.S. missiles on the moon.

The Outer Space Treaty would outlaw both the annexation of the Moon and its use for aggressive purposes. As enshrined in Article IV of the OST, States Parties undertake not to install nuclear weapons or any other kinds of weapons of mass destruction on celestial bodies, not to establish military bases, installations and fortifications thereon, not to test any type of weapons and to conduct military manoeuvres on celestial bodies, and vow to use the moon and other celestial bodies exclusively for peaceful purposes.

We made this parenthesis in order to point out that, initially, sovereignty on the Moon was not contemplated for its mineral riches, but for its military potential. With the decline of its strategic importance and also with the non-aggressive tenets of the Outer Space Treaty, the case can be made for the abrogation of the non-appropriation principle in its "sovereignty over natural resources" context, provided the non-aggression tenets are still valid.

Is a withdrawal from the OST possible? Article XVI of the document authorizes any State Party to denounce the Treaty by giving a one year notice, whereas Article 20 of the Moon Agreement contains a similar provision. This approach would be a radical paradigm shift from the past non-appropriation policy officially professed by most of the nations.

As early as a year after the entry into force of the OST, David S. Myers (1975, p. 69) acknowledged that the withdrawal provision could be used, "sometimes in the future", in furthering the needs of the "technologically superior states". Lamenting the likelihood that States will be unable to "mine or secure resources in space ... without coveting advantages" for themselves, and conceding that the differences in technological progress among the principal competitors may result in a lesser cooperation in outer space matters, he doubted that the *res communis* principle could be successfully applied over a period of time. Myers envisaged nonetheless the likelihood that a State may "prompt areas or resources effectively to the exclusion of others" without making formal claims of territorial sovereignty, this situation allowing a State to gain an important economic advantage whilst "reap[ing] political benefit" by not claiming sovereignty over that celestial area.

In the previous chapters, it has been shown the importance of sovereignty in the context of policing property rights. Abrogating the non-appropriation principle seems to clarify once and for all the "paper conflict" on space property rights, a State having in that case the sovereign right to regulate the mineral resources as its municipal law would become the *lex situs*. Were the non-appropriation clause of the Outer Space Treaty denounced, this would revert the extraterrestrial realms to the status of *res nullius* – open for occupation (*res nullius fit primi occupantis*). States would be permitted to acquire sovereignty on celestial bodies, being able to subsequently "license, grant, permit, recognise, authorise and/or enforce rights of private ownership and development of such assets" (Sterns et al., 1996, p. 53). According to Christol (1997, p. 39), the absence of sovereignty in the extraterrestrial realms prevents a State from granting property rights over resources. "Only the pre-existing condition of sovereignty" – writes he – "can sustain a permanent national claim to property or a permanent allocation of property rights".

Pursuant to Article 2.1 of the Charter of Economic Rights and Duties of States (UN General Assembly Resolution 3281 (XXIX) of 12 December 1974), "[e]very State has and shall freely exercise full permanent sovereignty, including possession, use and disposal, over all its wealth, natural resources and economic activities." Whereas some authors and activists envisage a sovereign with a protective role, others wish for a more pro-active role of the State in allocating property rights. Should appropriation be performed through acts of governmental agents, on the plane of property rights this would confer the extraterrestrial lands the character of national "public lands"; should the seizure be performed by private persons, their State of citizenship would be able to protect their interests.

A number of authors would favour an international sharing of the extraterrestrial realms. Barbara Heim (1990, p. 846) wrote that one possible option of solving the extraterrestrial ownership issue is:

> to divide ... the moon into distinct geographic portions, with each nation getting a share. Those nations possessing the technology could exploit their portions today; developing nations would reserve their region for future development.

However, due to foreseeable conflicts regarding the dimensions and locations of the national shares, a "pencil-on-the-map" division of the celestial bodies into national portions would be difficult to implement. Conflicts would arise also in the matter of attributing scarce resources, such as the crater containing what seems to be the only trace of water on the Moon. In 1962, science-fiction author Hugo Gernsback suggested in a *bona fide* article a means of dividing the Moon among the nations who send their citizens to become temporary workers on the natural satellite – a more practical approach than the one previously presented. According to him, these countries may be entitled to wedge-shaped lunar slices – running from pole to pole, from the center of the Moon to the surface – in proportion to the number of space workers they send. Gernsback considered his method – taking into account the fact that "lunar real estate should not be measured only by the surface but by the depth, too" – as allowing each nation in cause to "exploit its entire slice without encroaching on anyone else's territory" (Gernsback, 1962, pp. 4–5).

A division of the geostationary orbit based on a "wedge" approach – albeit on a pure geographical rationale – has been actually attempted by some nations. On December 3rd, 1976, the representatives of eight Equatorial States met in Bogota, Colombia and signed the "Declaration of the First Meeting of Equatorial Countries". In the document, the delegates of Brazil, Colombia, Congo, Ecuador, Indonesia,

Kenya, Uganda and Zaire expressed their dissatisfaction with the 1967 Outer Space Treaty. In their view, the said document "cannot be considered as a final answer to the problem of the exploration and use of outer space" given that the developing countries "were ... not able to observe and evaluate the omissions, contradictions and consequences of the proposals which were prepared with great ability by the industrialized powers for their own benefit" (para. 4). Considering that the solutions proposed by the International Telecommunications Union are "at present impracticable and unfair" and that "the geostationary orbit and the frequencies have been used in a way that does not allow the equitable access of the developing countries that do not have the technical and financial means that the great powers have", the delegates considered being "imperative for the equatorial countries to exercise their sovereignty over the corresponding segments of the geostationary orbit" (para. 1). Consequently, the participants declared that, as a "physical fact linked to the reality of our planet, [...] the segments of geostationary synchronous orbit are part of the territory over which Equatorial states exercise their national sovereignty", having "decided to proclaim and defend on behalf of their peoples, the existence of their sovereignty over this natural resource" (para. 1). In their view, this approach would not be illegal, given that the lack of definition of outer space in the OST "implies that Article II should not apply to geostationary orbit and therefore does not affect the right of the equatorial states that have already ratified the Treaty" (para. 4). Confronted with the lack of international recognition and with the inability to enforce their orbital claims, most of the signatories abandoned their 1976 rhetoric (Golrounia and Bahrami, 1997, p. 305). Colombia, however, did not renounce its extraterrestrial claim; its 1991 Constitution proclaims that "the segment of the geostationary orbit, the electromagnetic spectrum and the space in which it operates" is a part of its national territory, "in accordance with international law or the laws of Colombia in the absence of international regulations" (Article 101). The next article states that "[t]he territory, together with the public resources that are part of it, belong to the nation".

A number of American authors and organizations favour a "go it alone" approach. Territorial expansion and homesteading have long been a part of the American ethos; from 13 colonies on the Atlantic coast in 1776, the United States have evolved, in less than two centuries, into a world superpower. It is, therefore, not surprising that a number of voices suggest carrying the "Manifest Destiny" and the "Frontier Spirit" into outer space and transforming the Moon into the 51st State. In the 1990s, the space property rights issue took a step forward through its active endorsement by space advocacy groups militating for a proper homesteading of the celestial bodies.

In February 1992, the California Space Development Council – a coalition of grassroots, pro-space activist and educational organizations in California and Nevada – passed a resolution titled "Luna 2010" (also known as "Option 51"). In the document, the organisation recommended that the United States should seek to amend the Outer Space Treaty in order to allow for national claims of sovereignty over celestial bodies. Should the treaty not be amended within five years, the CSDC advocated US withdrawal from the Treaty effective December 31st, 1998. If this occurred, the non-appropriation principle would no longer be in force for the United States, and the group argued that the US should claim the northern hemisphere of the Moon and develop it as United States territory, with the aim of "constructing a settlement of at least 100 people on the Moon by the year 2010". The CSDC envisaged the development of the territory into one or more new states of the US. In the area of space property rights, the proposal advocated, echoing the 1862 Homestead Act, that:

> The northern hemisphere of the Moon shall be made available to homesteading by United States citizens, and other groups that congress may approve. Any homesteader surviving for six months on the Moon shall be granted title to 100 square miles of lunar land (CSDC, 1992).

The National Space Society – America's pre-eminent citizen's voice on space – considers the lack of sovereignty in space as having "left a void of any legal system that would enable private entrepreneurs and companies to plan and execute commercial space activities on the Moon and other celestial bodies". This state of fact is seen as a political barrier to space settlement, given the NSS's opinion that:

> space development and settlement will occur most efficiently, and humanity's survival and growth will be best ensured, if every human being is allowed the opportunity to own property in space and/or on other world (National Space Society, 1999).

One of the NSS leaders, Alan Wasser, would take one step further by drafting a full "Space Settlement Prize Act", based on his "Space Settlement Initiative", whereby the US courts and agencies would recognize, certify and fully support land ownership claims based on use and occupation:

> for any private entity which has, in fact, established a permanently inhabited settlement on the Moon, Mars, or an asteroid, with regular transportation between the settlement and the Earth open to any paying passenger (Wasser, 2004a).

In 2004, Sam Dinkin, a space pundit, economist and investor, urged the US Department of State to "think outside the sphere" and imagine "a new Outer Space Treaty that supports property rights and individual and corporate rights in space". This would follow an already established pattern:

> If we can bust the ABM treaty, stick our tongues out at the International Criminal Court, step on the landmine treaty and blow smoke at the Kyoto Treaty, let's get some bang for our buck and not stop there. Let's withdraw from the Outer Space Treaty and establish a private property rights regime that opens up a new land rush into space (Dinkin, 2004).

Space commentator Rand Simberg (2004) envisages a new space age built on the ashes of the old one; a space age based on the "traditional values that opened up earlier frontiers – individualism, free enterprise, daring, and liberty". In order to ensure the success of the new space vision, he calls the US Government to take several steps; one of them is renegotiation or withdrawal from the OST due to its banning of national sovereignty that makes the defence of private property rights in space "problematic". In Simberg's view, the OST is, like the old space age, a relic of the Cold War, adopted in the 1960s in the "full flower of the mood" of decolonization and socialism in order to prevent a genuine space race and "save funds for 'terrestrial' needs".

Another space activist, Kevin C. Davis (2005), maintains the "Anti 1967 UN Space Treaty Page", aimed at starting a movement to have the US to withdraw from the said treaty "[b]asically for one reason and that reason is property rights in space". The OST is seed by Davis as "bad because it does not allow people to claim an asteroid or allow people to claim some land on the Moon and Mars. Basically it prohibits private development in space".

6.2.5 Private Appropriation Under an Independent Regime

Once permanent settlements have been established in outer space and celestial bodies, the issue of their self-administration and independence could arise. On the plane of property rights, by acquiring statehood the extraterrestrial realms would be entitled to have the appropriate *lex situs* protecting ownership of real estate located outside Earth's atmosphere. The issue of independence of space settlements raises several legal and ethical questions. Should an inhabited Moon or Mars be subject to the Earth colonialism, or be granted home rule? Would the future lunar colonists be entitled to "nationalise" what is considered to belong to Mankind as a whole? Is the "Boston Tea Party" going to be repeated in a "Lunar Helium-3 Party"?

According to Ernst Fasan, the exercise of jurisdiction and control by the state of registry over a settlement on Mars may become difficult, even impractical or impossible. With the passage of time and with the attainment of self-sufficiency, the settlement would create its own rules and "become a new nation in itself" (Fasan, 1994, p. 53). If the inhabitants of a station decide to act as a community and if the State of registry has loosened its control, then, according to Fasan (1994, p. 55), the chief conditions for the emergence of a new nation may be met. As to the non-appropriation of celestial bodies, the *fait accompli* will prevail:

> facts will have normative power. An area on the celestial body will be inhabited; its minerals will be used; inhabitants will move through such territory, and the "rules of conduct" will be valid within such an area (Fasan, 1994, p. 54).

As maintained by the authors of a study on lunar Helium-3 mining, this resource is likely to make lunar settlements "self-sufficient, economically attractive, and self-governing" (Bilder et al., 1989, p. 74), hence a regime for the development of lunar resources should provide for an eventual self determination for the settlers once the level of population, economic viability and political power make such a prospect feasible (ibid., p. 77). Patricia M. Sterns and Leslie I. Tennen (1979, p. 221) called for achieving settlement competence by an "international agreement of recognition and capacity" granting a "limited form of home rule" to communities in outer space. This is because the *fundator terrani* – that is, the terrestrial entity which established the space settlement – is unable to realistically administer the day-to-day operation of a settlement, whereas inhabitants of a settlement are better equipped at handling local matters (ibid., p. 223).

As early as 1957 Andrew G Haley (quoted by Quadri, 1959, p. 595) considered that the Moon shall be declared "*territoire libre, independant et autonome*" But – comments Rolando Quadri – the Moon does not have any inhabitants! While a number of authors consider a lunar autonomy once permanent settlements are in place, other scholars and activists call for an interim governance system, prior to the actual settlement of the extraterrestrial realms – such as the "Regency of the United Societies in Space", presented in a previous chapter. The past century witnessed a number of utopian attempts at creating a space governance regime – as seen *supra* from the account of the Nation of Celestial Space. This trend is continuing into the new millennium. The "Lunar Republic Society" managed by David Ferrell Jackson, maintains an "independent and sovereign Republic of Luna", established on the occasion of the thirtieth anniversary of humankind's first visit to the Moon. In 2001, Dennis Hope announced as well that the Lunar Embassy, in accordance with the wishes of its more than 1 million "property owners" at the time, was forming its own government. In March 2004, the Constitution of the Galactic Government was presented online for ratification by all "property owners", and the results were a 99.7% vote in favour of ratification. "The Galactic Government is now a fully ratified sovereign democratic republic nation" – boasted Hope in 1995. As such, it contacted all governments on Earth, with the purpose to set in motion the opportunity for recognition of their government through diplomatic channels. "To date we have heard from 77 governments. We have received positive responses from 49 of them and 'don't bother us' responses from 28", disclosed Hope. "Our goal is to eventually be recognized by all governments on Earth. No doubt that will take some time." (Dennis Hope, personal communication, December 1, 2005).

The issue of "micronations" in outer space – unrecognized fantasy constructs unable to physically control the territories they claim – is as trivial as the matter of "extraterrestrial unreal estate", examined in the first chapter. The issue of independent or autonomous space settlements is, on the contrary, a genuine item. We have to agree with the opinion expressed by Aldo Armando Cocca at the dawn of the space age:

> The Moon cannot be declared independent by the States of the Earth. ... In the event of an effective occupation and supposed 'colonization' of the Moon, the power to declare itself independent of the States of the Earth could only be derived from the 'colonists' who could vote to sever all ties of political dependence from the States of the Earth (Cocca, 1958, p. 35).

6.3 The Acquisition of Ownership in Outer Space

We have shown in the first chapter that, in order to acquire possession, one must prove both an intention to take a good – *animus* – and the performance of a physical act giving effect to that intention – *corpus*. While the "celestial property owners" featured at the beginning of this work base their claims on pure *animus*, several space exploration companies intend to claim ownership of space resources – namely, asteroids – only after having established effective possession by means of robotic prospectors. Can title to space property (should space assets be claimable) be gained this way?

Advocates of telepossession – i.e., property acquisition through telepresence, appeal to the 1989 case of *Columbus-America Discovery Group, Inc. v. The Unidentified, Wrecked and Abandoned Sailing Vessel, S.S. Central America*. In that case, a U.S. District Court in Virginia ruled that a salvage firm could claim an undersea wreck it had explored via "telepresence" with an undersea probe equipped with a TV camera. Human presence on the ocean floor was not required to claim the wreck; instead, the court allowed telepossession, defined as:

> (i) locating the object searched; (ii) real time imaging of the object; (iii) placement or capability to place teleoperated or robotic manipulators on or near the object, capable of manipulating it as directed by human beings exercising control from the surface; (iv) present intent to control (including deliberately not disturbing) the location of the object.

According to Richard Westfall (2003), establishing telepossession of resources in space should involve three tasks:

- **Telepresence** (visual observation of the site) – telepossessor must be able to show live video pictures of the resource site;
- **Telemetry** (communication with and knowledge of the location of the site) – telepossessor must know where the resource is and be able to communicate with the equipment thereon; and
- **Telerobotics** (manipulation of the resources at the site) – telepossessor must demonstrate ability to manipulate the resource site presumably (but not necessarily) through purifying the ore, reducing metals, or manufacturing parts *in situ*.

An interesting precedent justifying the acquisition of property through telepossession can be found in the Roman law, where a person could acquire ownership or other rights not only by his own direct actions, but also through the actions of slaves under his power (*in potestate*) (Roby, 1902, p. 432). Slaves has a similar legal status as modern robotic space probes; the word "robot" itself was derived from the Czech "robota" – meaning "labour" in the 1920s by the author Karel Čapek (Zunt, 1998). Roman slaves were mere instruments for their master (Roby, 1902, p. 452), being nothing more than chattels, objects of property rights and having no independent legal existence as persons. Their master had the property in anything that the slaves acquired (Ibid., p. 53). In the absence of a statement to the contrary, a slave owned in common by two or more masters acquired for them in the ratio of their respective shares in him (Ibid., p. 433).

The idea of claiming space assets through telepresence originated with Jim Benson, the founder of SpaceDev. In 1988, Benson stated his intention to land a probe on an asteroid and, besides collecting scientific data, to plant his corporate flag there (Dixon, 1998, p. 29). The search for candidate asteroids started in late 1996 – early 1997 (Clark, 2000); in 1998, Benson declared the target to be Near Earth Asteroid 4660 Nereus, planning to launch Near Earth Asteroid Prospector

(NEAP) in 2001 and rendezvous with Nereus in May 2002 (Benson, 1998). Funding problems have delayed Benson's mission and, by 2000, he abandoned Nereus as a target – making his decision in the light of the ongoing discovery of Near Earth Asteroids, which offer a number of better targets. Asked whether he would "declare war" if he claims an asteroid and then somebody lands on it, Benson declared that "the best thing to do would be to welcome them and offer hospitality", but "[i]t would be fun to charge someone a dollar for landing rights, though" (Clark, 2000). The intention behind Benson's plan was "to set an important and historic precedent and to initiate a global, public discussion of private property rights in space". In his view the landing of a privately financed robotic representative, with its resource assessment instruments, is sufficient to make an ownership claim. Benson stated that there is no entity on Earth with standing in space and there is no appropriate body to which it could make its claim, the only possible option being to make the claim to the public. In his view, SpaceDev's actions "will force the issue of private property rights in space into the public arena, to be decided by public opinion" (Benson, 1998, pp. 46–47).

Benson is not the only entrepreneur who seeks to appropriate an asteroid by means of remote-controlled *pedis possessio*. Galactic Mining Industries, Inc, a US company, intends to send telepossession probe sets to Near Earth Asteroid 2004 GU9, using use Gamma and Neutron Spectroscopy to assay the asteroid and claim ownership of the said resource, with the view to allow for the leveraging of financing for the development of this space resource and establishing thus a precedent for all other corporations around the world (Galactic Mining Industries, 2006). Two variants of telepossession are suggested by this company.

The first method envisaged by Galactic Mining Industries is through "Telepossession Probes of Galactic Mining" performing the tasks of a human and establishing thus title. Such a probe would consist of a lander and a relay spacecraft, which would rendezvous with an asteroid in space. The lander would perform tasks such as assaying the composition by various chemical and radiological methods; manipulating the asteroid by landing and hard-mounting to its surface, and by collecting samples; beneficiating the resource by taking some of the material thereon and converting it to a product, – and sending telemetry to the relay spacecraft and the operators of the probe on Earth. The relay spacecraft would perform the tasks of observation and remote sensing of the resource, as well as telemetry from the Earth to the lander and from the lander to Earth (Westfall et al., 2003).

The alternative method involves tagging the target asteroid with a RADAR Transponder package which, when interrogated, would send a string of data with unique identifiers imbedded and would also assist in the subsequent docking of automatic development payloads. Gary Rodriguez, the promoter of the method, suggests that before one's registration can be approved, a telepossessor should be required to maintain continuous transponder operation for ninety days, being allowed three years to replace a malfunctioning transponder (Rodriguez, 2004). According to him, this is "[a] relatively inexpensive claim mechanism, but a claim nonetheless", given that "entire terrestrial continents have been claimed based upon a mere piece of cloth affixed to a stick" (Westfall et al., 2003).

Should the plans of SpaceDev and Galactic Mining Industries be eventually put into practice, we believe that such entities may well have a legal standing to make a property claim in the presence of both animus and corpus. Of course, the validity of ownership acquisition through space telepossession is subject to the permissibility of fee simple property rights in the extraterrestrial realms. Space lawyer Glenn H. Reynolds agrees that claiming property by landing a robot may well be possible; he nonetheless considers that the size of the claim has to be reasonable – e.g., one can claim "the quarter of a square mile in which your robot rolled around", but not a whole planet (Scripps Howard News Service, 1998).

6.4 Arguments for the Frontier Paradigm

The homesteading movement in the United States received momentum from the fact that it would promote the settlement of the High Plains, regarded until 1862 as the "Great American Desert" (Robbins, 1942, Chapter XIV). A century later, Apollo 8 astronaut Frank Borman would dub the Moon an "expanse of nothing" (Chaikin, 1994, p. 121). The homesteading paradigm is likely to transform the lunar desert in the same manner as it transformed the 19th Century United States.

Most of the arguments used for defending property rights on earth are also of currency in the extraterrestrial realms, and some will be presented in the next chapter when criticizing the common heritage of mankind model. Support for a legal regime favouring a privatization of the celestial bodies comes from the right side of the political spectrum. Conservative and libertarian authors promote:

- individualism instead of collectivism;
- nationalism before cosmopolitanism;
- competition rather than cooperation;
- the economic interests of the haves before those of the have-nots;
- economic liberty, i.e. free market, instead of economic equality;
- acceptance of inequalities instead of redistribution of wealth;
- efficiency as opposed to equity, or fair process as opposed to fair outcome;
- laissez-faire versus interventionist economics;
- smaller versus bigger government.

A first argument is more practical than ideological. Most of the resources of outer space are unlimited; it is estimated that the Oort cloud contains trillions of comets. In 1968, Arthur C. Clarke and Stanley Kubrick, the creators of "2001:A Space Odyssey", calculate in a wonderful sample of astro-demographic mathematics that:

> Since the dawn of time, roughly a hundred billion human beings have walked the planet Earth ... [B]y a curious coincidence there are approximately a hundred billion stars in ... the Milky Way. So for every man who has ever lived, in this universe, there shines a ... sun. ... And many ... of those alien suns have planets circling them. So almost certainly there is enough land in the sky to give every member of the human species, back to the first ape-man, his own private, world-sized heaven – or hell (Clarke and Kubrick, 1999, p. 7).

There is more lunar and asteroidal material than water in Earth's oceans, which is considered, in practical terms, an unlimited resource. Given the abundance of extraterrestrial resources, it would be nonsensical to forbid their private appropriation.

A number of authors justify space property rights as a necessary engine for pushing forward the development of the extraterrestrial realms. In 1966, Argentine scholar Enrique Ferreira (quoted in Ferrer, 1969, p. 146) saw no reason to ignore property rights in outer space, seen as resulting from the scientific and economic effort spent in reaching and dominating the extraterrestrial realms. In his view, everything acquired therefrom would benefit all of humankind, directly or indirectly, as any other discovery or invention. Ferreira considered that the denial of the above would "paralyze from the very beginning, any individual or national effort, which would prejudice everyone and benefit nobody". Stephen Gorove (1983) believes that the basic tenets of the Moon Agreement – the common heritage, the equitable sharing and the international regime – introduce legal uncertainties of such a magnitude as to diminish or even halt the ideas and efforts of the private enterprise to invest billions of dollars in an endeavour already risky through its pioneering nature and the adversity of the space environment.

Roberts et al. (1996) believe that the future of lunar exploration depends upon a "cohesive system of property allocation and rights", the current absence of specificity on this regard making "untenable" the commercial exploitation of these resources. "What financier would invest the billions of dollars necessary for lunar operations without the security of clear rights to the resources being recovered?" – ask the three authors. Certainly not James William Benson, founder and chairman of world's first commercial space exploration and development company SpaceDev, who considers the quick establishment of strong private property rights in space as serving the best interest of humanity, these being "essential to the rapid and orderly development of space". Thus, "[i]nvestors need to know that their risk money can be rewarded through the exploration, discovery, ownership and utilization" of space resources, being "more willing to invest in activities that can be profitable, and ... more comfortable operating in a known and stable business environment" with property rights assured (Benson, 1998, pp. 46–49).

Karen Cramer (1997, p. 352) considers that some form of extraterrestrial land use regulation is needed in order to prevent interference among potential users of lunar territory, as the absence of property rights would breed anarchy, hindering large investments of capital. In Alan Wasser's (2004b) view, space property rights are "only a means to an end, not an end in itself". Their real function is to facilitate the human expansion into space by offering a financial reward for privately funded settlement, being "the only way to create an economic incentive sufficient to encourage private investment to develop affordable human transport to the Moon and Mars". While they may not be enough, given the lack of knowledge regarding "how much it will cost to develop safe, reliable, affordable space transport, or how long it will take", Wasser is confident that the promise of property rights for space settlement would help generate the investment needed for developing affordable space transport, being "a very low cost, low risk, 'do-able' way to attract that supplementary venture capital".

Glenn Harlan Reynolds (1995, pp. 118–120) believes that lunar resources will not be developed without an appropriate legal regime, and that the one most desirable and most likely of being accepted by the spacefaring nations would be founded on private property rights with minimal governance. He hopes that the spacefaring nations will soon seize the opportunity to work out such a system, while acknowledging that a unilateral regime, built on the domestic law of individual nations, could provide the needed protection for property rights, at least in the first phase of lunar resource development. Reynolds nonetheless fears that the latter approach would increase the potential for conflict among States, and it appears to lack "political legitimacy in the world community"; he acknowledges that the confidence and stability that such a regime is designed to create might be undermined by the fear of hostile action by other nations – be it diplomatic, military or economic. These hindrances are not insurmountable in his view, due *inter alia* to the precedent represented by the 1980 US Deep Seabed Hard Mineral Resources Act which established a unilateral system for governmental licensing of seabed mining, yet he recommends using unilateralism only as a last resort due to the problems associated with it.

Perhaps the best arguments for the privatisation of the extraterrestrial realms has been brought by John Locke (1690, Section 37) more than three centuries ago. In his view – to which we adhere – privatization enhances the common heritage, provided there is still enough left for the others:

> [M]en had a right to appropriate, by their labour ... as much of the things of nature, as he could use: yet this could not be much, nor to the prejudice of others, where the same plenty was still left to those who would use the same industry.

Locke did not justify greed, bearing in mind the interests of the fellow men. Yet, as shown *infra*, this interest cannot be the entitlement to the work of the other, but the right to use one's "industry" for homesteading the common heritage. In his correct reasoning, he who proceeds at appropriating land by the means of his labour "does not lessen, but increase the common stock of mankind", because enclosing an acre and using it yields more essential products than if the land were left "waste in common". The privatization is seen, thus, as an active administration of the public trust stemming from the common heritage in its original sense:

> And therefore he that incloses land, and has a greater plenty of the conveniences of life from ten acres, than he could have from an hundred left to nature, may truly be said to give ninety acres to mankind.

The support for property rights in outer space ought to come with a caveat. It is hereby offered that private property, even in its terrestrial form, is an eroded concept. Throughout the last century, the "owner" has involved into a mere "privileged user" of a certain object. The State taxes that object and the benefits derived thereon, and can confiscate that object by means of eminent domain. Fee simple ownership, as an epitome of property, has been drawn into extinction by the gradual advance of the society into private matters. It is therefore essential to bear in mind that, even if formally allowed, landed private property rights in outer space would still be subject to taxation, eminent domain and other State powers. Support for property rights in outer space needs to be complemented by support for property rights on earth.

6.5 Arguments Against the Frontier Paradigm

Whereas most of the ideological arguments brought forth against privatizing the extraterrestrial *res publica* will be presented in the next chapter, a number of authors expressly criticize the frontierist ethos. Clas G. Wihlborg and Per Magnus Wijkman (quoted by White, 1985, pp. 38–39) deem that a squatter's rights regime would echo the land and gold rush that occurred during the westward advance of the American civilization. With the extraterrestrial realms seen as a new frontier, they fear that the best equipped enterprises in the most technologically advanced countries would haste to claim the most valuable orbital slots and frequencies, with little of the value being left for other nations. The two authors oppose the frontier paradigm on the ground that "space is the common property of mankind", and, given the disparity between the developed and the developing world, such a regime would be inequitable for the latter, being "unacceptable to most countries".

Another opponent is Alan Marshall (1995, p. 46), who adopts a politically correct rhetoric in criticizing the frontier approach as an expression of 19th Century bourgeois economic policy tied with populist ideology. In his view, frontierism is ultimately an economic philosophy, an embodiment of "capitalist imperialism", since it entails the seizure of resources seen as previously unowned, and holds nothing of value in the frontier real estate except what can be converted into capital. By transforming barren land into an economic resource, the frontier is seen as contributing to the wealth of the bourgeoisie more so than to that of the nation. Marshall considers that the outlook of those who advocate pioneering the space frontier resembles the attitude of the "imperialistic capitalists" under which the West succumbed. As these "arrogantly ignored" the value of the landscape and that of the native peoples, so the supporters of space frontierism consider the celestial bodies of the solar system as "valueless hunks of rock until acted upon by humans to produce economic value and contribute to capital accumulation".

The abrogation of the non-appropriation principle – or of the OST as a whole – is opposed by some authors on the grounds that this would open a Pandora's Box. As early as 1959, the United Nations *Ad Hoc* COPUOS stated its fear that "serious problems could arise if States claimed, on one ground or another, exclusive rights over all or part of a celestial body" (Dembling, 1997, p. 35). A supporter of space property rights, Michael J. Listner (2005) warns nonetheless that withdrawing from the OST – the "fastest and most efficient solution" – would create significant political fallout and would not be "palatable". Another option – that of amending the OST in order to work around the *res communis* doctrine and to make it more "user friendly" – is likely to lack, in Listner's view, the required support among the signatories to the Treaty.

In a similar line of reasoning, Joanne I. Gabrynowicz (2005, pp. 30–31) believes that space law, in its current incarnation, is not outdated, and that withdrawing from the OST is a wrong pitch. She laments that the whole array of commercial interests has been reduced to a single issue in the bipolar debate surrounding the existence of property rights in the OST regime. Gabrynowicz deems that the document in cause, being silent on the issue of property rights, contains no prohibition in this field. Rejecting the OST on the grounds of its silence on property rights will bringing into question – according to her – the rest of the regime containing the basic legal structure necessary for commercial activities in outer space, and the applicability of the private law developed throughout the years in the form of insurance agreements and contracts. In her view, the OST regime supports the American interests during the current era of globalization, and is considered as binding customary law even for the States who are not parties to it. As a solution, she suggests instead developing the political will to clarify the treaty regime in the field of property rights, using the amendment process. Another option advanced by Gabrynowicz is to establish municipal laws that clarify or fill in the legal gaps in the international regime – as already done in the matter of launches, commercial remote sensing, telecommunications, among other space activities.

The frontier paradigm has been abandoned in the United States in 1976, when the Congress mandated the Bureau of Land Management to generally retain the public land in public ownership. Parts of the American *res publica* – lands that have been identified by the Federal Government as unneeded or that would be better utilized in private ownership – are still made available for sale at a fair market value (US Bureau of Land Management, 2000), though disposal of land is nowadays the exception rather than the rule.

6.6 Conclusion

The frontier paradigm has proven its worth on our planet, and it most likely will do so in the extraterrestrial realms. Homesteading is likely to transform the lunar desert in the same manner as it transformed the 19th Century United States. Space is indeed a new frontier calling for individualism rather than collectivism, and its challenges need to be addressed with a legal regime favourable to property rights. Such a regime is seen by many authors as not only useful, but also as the only means of opening the extraterrestrial realms to settlement, given the reluctance of most industrialists to invest money in an endeavour without having the security that they will enjoy the benefits. It may also occur that a minority of investors, with a bigger tolerance to risk, would adopt an anarcho-capitalist approach and "cross the Alleghenies" without backing from a sovereign State.

Given the abundance of extraterrestrial resources, it would be nonsensical to forbid their private appropriation. Securing property rights would be a small price to pay, and more beneficial to humankind, compared to the alternative of keeping the extraterrestrial realms undeveloped. The practical arguments against the Frontier paradigm may have merit, but the issues raised can be tackled. The ideological arguments, nonetheless, are emotional rather than rational.

Whereas the frontier paradigm is outlawed in the current incarnation of the international law of outer space, law is a dynamic phenomenon and it may evolve towards a regime supportive of property rights in outer space. A shift from the *res publica* approach may be in the cards, given the official support of the Aldridge commission for property rights. Until this shift happens, the non-appropriation principle remains nonetheless the *lex lata* in the extraterrestrial realms.

Chapter 7
The Common Heritage of Mankind: Reaping Without Sowing

From each according to his abilities, to each according to his needs

Karl Marx (1875)

7.1 Introduction

A spectre is haunting outer space: the spectre of communism. "In the earlier epochs of history" – wrote Marx and Engels (1848, para. 1) – "we find almost everywhere a complicated arrangement of society into various orders ... In ancient Rome we have patricians, knights, plebeians, slaves; in the Middle Ages, feudal lords, vassals, guild-masters, journeymen, apprentices, serfs". And in the space age, we have the antagonistic spacefaring and non-spacefaring nations, developed and developing states, the "North" and the "South".

The post-Sputnik and post-colonial era could not escape the materialist conception of history, whereas the world and its extraterrestrial surroundings are the scene of the class struggle between the haves and have-nots. Marx, Engels and many other communists sought an end to capitalism and the establishment of an egalitarian society. The 20th century saw the establishment of several Communist States, and the adoption of several Socialist principles into the mainstream. In the 1970s, the Marxist ideas received escape velocity through the work of the United Nations, who oversaw the drafting of the "Agreement Governing the Activities of States on the Moon and Other Celestial Bodies". Of special importance are the provisions of its Article 11.1 stating that "[t]he moon and its natural resources are the common heritage of mankind", of Article 11.5 whereby States Parties agree to create "an international regime, including appropriate procedures, to govern the exploitation of the natural resources of the moon", and of Article 11.7.d, where one of the main purposes of the above regime is revealed as being:

> An equitable sharing by all States Parties in the benefits derived from those resources, whereby the interests and needs of the developing countries, as well as the efforts of those countries which have contributed either directly or indirectly to the exploration of the moon, shall be given special consideration.

Whereas at the beginning of the Space Age the Soviet technological superiority sparked fears in the United States that "space would soon be *res communist*, not *res communis*" (McDougall, 1985, p. 202), two decades later the Moon Agreement

V. Pop, *Who Owns the Moon?*, Space Regulations Library Series 4, 121
DOI 10.1007/978-1-4020-9135-3_7, © Springer Science+Business Media B.V. 2009

effectively planted the Communist standard in the lunar soil. The following chapter will critically analyze and rebuke the concept of the Common Heritage of the Mankind (CHM), as envisaged in the Moon Agreement, as an archenemy to space development.

7.2 The Genesis of the CHM Concept

The Common Heritage of Mankind concept has been presented, in most of the literature, as having originated in the 1967 proposal of the Maltese Ambassador to the UN, Arvid Pardo, that the sea-bed and ocean floor beyond national jurisdiction should be declared the Common Heritage of the Mankind by the United Nations General Assembly (Paxson, 1993, p. 501). This is not wholly accurate, as both similar ideas and similar language have been used before.

During the Moon Agreement negotiations in the Legal Subcommittee of the UN-COPUOS, some national delegates were not pleased with the diplomatic term chosen, asking off-the-record – "Who died and left the moon to mankind?" (Thomas, 1980, p. 159). The same question was asked by Dekanozov (1974, p. 202):

> [I]f celestial bodies and their resources are common inheritance, and succession in civil law is transfer of the property of a decedent (testator) to his successor (heirs), who then should be considered as the testator and has he existed at all?

The heritage in question is not the result of a death, yet it stems from a testament – namely the Old Testament, where God blessed the first humans to "fill the earth and subdue it" (Gen 1:28, NIV). This tenet was seen by Pope Pius XII as referring not only to out planet, but to "the whole of creation which He offered for the human spirit to penetrate" (Cheng, 1957, p. 505).

The term itself has been used verbatim as early as the 1840s–1850s; as shown elsewhere in this work, US Senator William Henry Seward delivered a speech in 1850 where he described the newly acquired public domain as "part of the common heritage of mankind, bestowed upon them by the Creator of the universe", and called for its enjoyment "either in common or by partition" (Seward, 1850, paras. 22–23). The idea of a possible partition of the divine bequest is not new to Seward; in the 17th Century, John Locke suggested that "God gave the world to men in common; but . . . it cannot be supposed he meant it should always remain common" (Locke, 1690, Section 34). Communist editor Hermann Kriege (quoted by Marx and Engels, (1846) endorsed the promoters of the American land reform, who "call the soil the communal heritage of all mankind" and want "to preserve as the inalienable communal property of all mankind the 1,400 million acres of land which have not yet fallen into the hands of rapacious speculators". For this reason, he shared in the goal "to place 160 acres of American soil at the command of every farmer, from whatever country he may hail, so that he may feed himself", a plot "which they needed only to settle and make fruitful with the labour of their hands".

While "heritage" is the word chosen to figure in the English text of the Moon Agreement, the Spanish text contains the term "patrimonio" and the French one refers to "patrimoine"; together with the Arabic, Chinese and Russian texts, these are equally authentic, according to Article 21 of the Moon Agreement. On an international level, calls for the international community to regard as "common patrimony" those "things which cannot be held by one nation without detriment to the others" were put forward, as early as the 1830s, by Latin American jurist Andrés Bello. In 1898, French lawyer Albert Geouffre de Lapradelle argued that the oceans should be considered "the patrimony of humankind", a concept shared six decades later by Prince Wan Waithayakon of Thailand, the President of the first Law of the Seas Conference (Schmidt, 1989, p. 23). In 1949, R.A. Smith wrote that "the Moon is not the property of any state; if it is untenanted, it is the common heritage of mankind", whereas Lionel Laming and Oscar Schachter declared, roundabout the same time, that not only the Earth, but all the solar system deserves to be considered as the heritage of mankind (Doyle, 1997, p. 10). Not much later, in a 1953 article, Joseph Kroell (quoted in Cocca, 1986, p. 17) stated that "L'espace extra-terrestre ... forme le patrimoine commun de l'Humanité". In July 1966, the United States were called upon by President Johnson to ensure that "the deep seas and ocean bottoms are, and remain, the legacy of all human beings" (Schmidt, 1989, p. 24). Five months later, the UN General Assembly adopted the Outer Space Treaty, whose Article I provides that "[t]he exploration and use of ... the ... celestial bodies ... shall be the province of all mankind". A resolution of the World Peace through Law conference adopted in early summer 1967 declared the seas beyond the continental shelf to be "the province of the UN" (Schmidt, 1989, p. 23), whereas in June of the same year, Aldo Armando Cocca introduced the CHM concept in the discussions being held in the Legal Subcommittee of the UNCOPUOS (Jasentuliyana, 1997, p. 58). Markus G. Schmidt (1989, p. 23) deems that, while many of the proposals above were "vague and without backing", Arvid Pardo's plan was "both timely and well conceived".

As suggested by Gennady M. Danilenko (1998, p. 250) the CHM concept needs to be assessed within the broader context of the developing world's efforts at that time to radically change the existing system of international economic relations. Indeed, the end of the 1960s saw a paradigm shift within the United Nations, with an increasing empowering of the developing countries; an important step would be taken in 1974, with the adoption by the UN General Assembly of Resolution 3201 titled "Declaration on the Establishment of a New International Economic Order". Elevating the concept of class struggle into outer space, this particular document outlined the widening gap between the developed and developing world, noted that "[t]he benefits of technological progress are not shared equitably by all members of the international community", and proclaimed the need for establishment of a "new international economic order based on equity ... interdependence, common interest and cooperation among all States ... which shall correct inequalities". In 1976, the Club of Rome released a report titled "Reshaping the International Order" (quoted in Vicas, 1980, pp. 301–302), where Outer Space was seen as "a geographical entity forming a 'common heritage of mankind" that calls for "ensuring that all nations, not only the powerful and rich, benefit from its exploration and exploitation" through its "effective management".

According to Danilenko (1998, p. 250), the developing world attempted to use the CHM concept during the negotiations for the 1982 United Nations Convention on the Law of the Sea and the 1979 Moon Agreement in order to establish a framework for exploiting the resources of the moon and of the sea-bed. The CHM would secure the direct participation of the developing world in the exploitation and international management of these resources, would prevent their monopolization by the developed countries where the developing world does not yet have the technology or financing necessary, and would distribute the benefits "primarily in the interests of the developing countries".

As it can be seen from the long history outlined above, the CHM paradigm underwent many changes – from a Creation open for homesteading and partition as seen by Locke and Seward – to its contemporary, egalitarian incarnation. The analysis will concentrate on its later form.

7.3 The Contents of the Common Heritage of Mankind

Whereas Stanley B. Rosenfield (quoted by Wiewiorowska, 1980, p. 84) deemed the CHM doctrine as being "without legal context [;] [a]ny meaning, and legal import, must be indicated by added principles", Article 11.1 of the MA is clear enough in stating that:

> the common heritage of mankind ... finds its expression in the provisions of this Agreement, in particular in paragraph 5 of this article.

Van Traa-Engelman (1980, p. 76) considers that the above tenet broke the "deadlock in the discussions on the applicability of the [CHM] doctrine as a legal concept", yet "the material content of the concept can still be subject to discussion". Indeed, as recognized by Andem (1999, p. 4), its provisions have been subject to conflicting interpretations. According to Francis Lyall (1998, p. 132), it is becoming accepted that, in any incarnation of it, the CHM concept presents a number of particular elements, namely:

> that certain regions should not be subject to national appropriation in any way, that there will be a management system for such an area, that the managers, be they state or international organisation, will act as representative of mankind,
>
> that any benefits from such areas will be shared internationally, and that the area will be used for peaceful purposes only.

Alan Marshall (1995, p. 52) agrees that the CHM is "not a well-defined concept", yet his "own predilections" suggest that the paradigm comprises the following traits:

1) non-appropriation (this is adequately encapsulated in the Moon Treaty, but is deficient in the Outer Space Treaty),
2) universality of applicability (to all states and to all parts of space, including space itself; this would thus make nations and private firms liable for rent payment with regard to orbital occupation),
3) universality of formulation (so that all states participate in the drawing up of space law),

4) equitable distribution of space resources (the meaning of equitable being decided by all states),
5) the use of space for peaceful purposes...

The vision of an economic system in outer space whereby property shall be controlled by a UN body and communally owned by all the nations is considered by Michele Hamilton (1998) as "nothing more than communism: a communal system of ownership where private property rights and free market principles simply do not exist". Indian scholar Mani (1996, p. 33) acknowledges the view of the developed countries, who regarded the CHM concept as being:

> the institutionalisation of the 'menace' of international socialism (read communism), proliferation of international bureaucracy, and threat to their technological superiority – in short ... an anathema to everything the Western capitalist ideology stood for.

The idea that the CHM is a socialist/communist concept is not far-fetched; indeed, both the legal analysis and the diplomatic statements support this view. Whereas several of the above characteristics do not pertain *ratione materiae* to the subject of this book, an analysis is needed as to the property implications of the Moon Agreement.

It is important to point out that the CHM as envisaged by the Moon Agreement is not the same with the CHM envisaged by the international law of the sea documents. As it will be shown in the next chapter, the UN General Assembly declared, in December 1969, a moratorium on the exploitation of the deep seabed resources pending the establishment of an international regime for the development thereon. No such moratorium has ever been proclaimed for the Moon. This is perhaps due to the general opinion that the exploitation of the manganese nodules of the deep seabed is more feasible with the current technology, whereas that of the lunar resources pertains to a more distant future – hence in a lesser need for regulation. The language of the Montego Bay Convention – as per its 1982 incarnation – is also tougher than the language of the MA. Article 137 of the UNCLOS, outlining the legal status of the "Area" and its resources, provides:

1. No State shall claim or exercise sovereignty or sovereign rights over any part of the Area or its resources, nor shall any State or natural or juridical person appropriate any part thereof. No such claim or exercise of sovereignty or sovereign rights nor such appropriation shall be recognized.
2. All rights in the resources of the Area are vested in mankind as a whole on whose behalf the Authority shall act. These resources are not subject to alienation. The minerals recovered from the Area, however, may only be alienated in accordance with this Part and the rules, regulations and procedures of the Authority.
3. No State or natural or juridical person shall claim, acquire or exercise rights with respect to the minerals recovered from the Area except in accordance with this Part. Otherwise, no such claim, acquisition or exercise of such rights shall be recognized.

The 1994 Agreement relating to the Implementation of Part XI of the Montego Bay Convention softens the hard-line provisions of the UNCLOS, given the "political and economic changes, including market-oriented approaches, affecting the implementation of Part XI" – as noted in its Preamble. The 1994 Agreement has a greater legal force than the Convention (Article 2.1). Whereas this is not the place for examining the legal status of the deep seabed or the oceanic incarnation of the CHM, the above example has been given as a means of illustrating the lack of uniformity – territorial and temporal – of this concept.

7.3.1 The Prohibition of Private Landed Property

As shown elsewhere in this work, Article II of the Outer Space Treaty prohibits the national appropriation of the extraterrestrial realms. While this denies, in practice, the private appropriation of the same, Article 11.3 of the Moon Agreement contains a facial prohibition of landed property in outer space:

> Neither the surface nor the subsurface of the moon, nor any part thereof or natural resources in place, shall become property of any State, international intergovernmental or non-governmental organization, national organization or non-governmental entity or of any natural person.

Libertarian commentator Edward L. Hudgins (1998) believes that the model for the Moon Agreement, where "[p]rivate property is explicitly banned", is the Soviet constitution. Indeed, whereas Article 13 of the 1977 Constitution of the USSR allowed Soviet citizens to own "articles of everyday use, personal consumption and convenience, the implements and other objects of a small-holding, a house, and earned savings", this did not apply to land. According to Article 11, "[t]he land, its minerals, waters, and forests are the exclusive property of the state", private citizens being provided, by means of Article 13, with a limited *jus utendi*:

> Citizens may be granted the use of plots of land, in the manner prescribed by law, for a subsidiary small-holding (including the keeping of livestock and poultry), for fruit and vegetable growing or for building an individual dwelling. Citizens are required to make rational use of the land allotted to them.

Article 10 placed the "socialist ownership of the means of production" at the "foundation of the economic system of USSR", whereas the "principal form of socialist property" in Soviet Union was State property, defined by Article 11 of the Soviet basic law as "the common property of the Soviet people".

Adding up Articles 11 and 13, it can be seen that all land and resources in the USSR was the common property of the Soviet people, while individuals could own a house, the implements of a farm, and could exercise user rights over the land granted to them for subsistence farming and dwelling. The Moon Agreement considers all land and resources on the Moon as the common heritage of the Mankind (Article 11.1), while individuals can own lunar stations, vehicles, equipment, facilities and installations (Article 12.1), and can exercise user rights over a limited area which is required for the needs of the station (Article 9.1).

Article 11 of the Moon Agreement is in resonance with the 1848 Manifesto of the Communist Party, whereby Karl Marx and Friedrich Engels called for the "[a]bolition of property in land and application of all rents of land to public purposes", entrusting the proletarians with the mission "to destroy all previous securities for, and insurances of, individual property". This is the core belief of Communism; as acknowledged in the same document:

> [T]he theory of the Communists may be summed up in the single sentence: Abolition of private property (Marx and Engels, 1848).

The private property in other legal categories – that is, extracted resources and planetary structures, as regulated by the MA, is scrutinized elsewhere in this volume.

7.3.2 The "Equitable" Sharing of *Jus Fruendi*

The key tenet of the CHM is the distribution of benefits. As an attribute of property, *jus fruendi* embodies the right to enjoy the income (fruits) derived from an asset. It will be seen elsewhere in the book that, even under the Moon Agreement regime, private property may exist in the resources extracted from the Moon. Nonetheless, this private property cannot be fully enjoyed, as Article 11.7.d of the MA provides for:

> [a]n equitable sharing by all States Parties in the benefits derived from [the natural resources of the Moon], whereby the interests and needs of the developing countries, as well as the efforts of those countries which have contributed either directly or indirectly to the exploration of the moon, shall be given special consideration.

In the 1979 report of the Independent Commission on International Development Issues, titled "North-South: A Program for Survival", the commission chaired by Willy Brandt (quoted in Vicas, 1980, p. 303) considered that "Global commons' is a neat catchword, but hardly appropriate", because:

> It connotes villagers in medieval England who have the right to pasture their cattle in the village commons. The space analogy is nations 'pasturing' their satellites in the global commons. The term connotes something of free access to outer space, but none of the distributional aspects of the 'common benefit' or 'common heritage'.

Indeed, whereas a *res communis* offers free access, it does not entail a share of the benefits. The villager pasturing his cow in the village commons needed not share the meat and milk with the other villagers, even if originated from the grass grazed from a common pool. In contrast, under the CHM regime, the lunar miner has to share with all other humans what his equipment extracted from the Moon. According to Stephen Gorove (1985, p. 227), as long as the space powers do not ratify the MA, their private enterprises are "entitled to acquire and retain space resources" for their own disposition "without limitation on possible profit".

Lenin (1917) defined Socialism as the "social ownership of the means of production and the distribution of products according to the work of the individual"; in his view, socialism will "ripen into Communism, whose banner bears the motto: 'From each according to his ability, to each according to his needs'. Christopher Pinto, Sri Lanka's ambassador to the Law of the Sea Conference (quoted in Danilenko, 1998, p. 257) expressed the view of the developing countries that the CHM, as incarnated in the Law of the Sea Convention, implements the principle according to which "each country will contribute according to its capacity and each will receive according to its needs". The tenets of the Moon Agreement hold middle ground between Socialism and Communism, providing for a share of the lunar benefits from each according to his ability, to some according to their work and to some according to their needs.

The CHM, as incarnated in the MA is even more radical than the Soviet doctrine. Article 18 of the 1918 RSFSR Constitution considered work as being the duty of every Russian citizen, proclaiming as its motto: "He shall not eat who does not work". Article 14.2 of the 1977 Soviet Constitution provides that:

> The state exercises control over the measure of labor and of consumption in accordance with the principle of socialism: 'From each according to his ability, to each according to his work'.

The 1918 Soviet contribution obliges every citizen to work; no part of *corpus juris spatialis* obliges any State to explore and use the Moon. Article I of the OST asks States to "facilitate and encourage international co-operation" in the scientific investigation of the extraterrestrial realms, whereas Article 4.2 of the MA expects that:

> [i]nternational co-operation in pursuance of this Agreement should be as wide as possible and may take place on a multilateral basis, on a bilateral basis or through international intergovernmental organizations.

Nonetheless, the exploration and use of the Moon is a freedom, not an obligation. Whereas Article 11.7 of the MA does give "special consideration" to the efforts of the countries having contributed to the exploration of the Moon, this is not the only basis for assigning *jus fruendi*; the "interests and needs of the developing countries" – whether lunar explorers or not – are part of the benefit distribution equation.

According to Brian M. Hoffstadt (1994, pp. 591–592), the meaning of "equitable" in the Moon Agreement creates uncertainties, especially as to its impact upon the amount of profit private companies will be allowed to keep; he deems that no investment will occur until "equitable" is defined and the profit distribution is settled.

Although not a Member State to the Moon Agreement, the US distributes globally some of the benefits from its space activities. As it will be shown *infra*, a number of lunar samples have been given, free of charge, to other countries, although their extraction was sponsored by the US tax payer. In the same time, NASA (2005) surrendered the copyright for still images, audio files and video, subject to certain conditions. This is done subject to a different benefits provision, where no distributive system is envisaged and no "regime" exists for distributing the same. Article I of the Outer Space Treaty provides that:

> The exploration and use of [the celestial realms] shall be carried out for the benefit and in the interests of all countries, irrespective of their degree of economic or scientific development, and shall be the province of all mankind.

Article 4.1 of the Moon Agreement repeats, in a different wording, the above clause, adding:

> Due regard shall be paid to the interests of present and future generations as well as to the need to promote higher standards of living and conditions of economic and social progress and development in accordance with the Charter of the United Nations.

State practice indicates that, under the terms of Article I of the Outer Space Treaty, States are willing to share the scientific results from space exploration, yet not the profits stemming from commercial space activities.

On December 13th, 1996, the UN General Assembly adopted a "Declaration on International Cooperation in the Exploration and Use of Outer Space for the Benefit and in the Interest of All States, Taking into Particular Account the Needs of Developing Countries" (A/RES/51/122). The Declaration evokes the provisions of the OST, without making any reference to the provisions of the Moon Agreement. An important principle is contained in para. 2 of the above document, allowing States the freedom to determine, on an equitable and mutually acceptable footing, all the facets of their participation in international cooperative ventures in the exploration and use of space. A special attention is paid to the contractual terms in cooperative ventures, which ought to be fair, reasonable, and fully conform to the legitimate interests and rights of the contracting parties. Whereas para. 3 calls countries with more advanced space capabilities to be particularly aware to the interest and benefit of countries with incipient space programmes and developing countries stemming from cooperative space activities conducted with them, no reference whatsoever is made to the distribution of profits from non-cooperative ventures.

According to Marietta Benkö, Schrogl (1997, p. 159), the above declaration denounced the "forced cooperation and dirigist approaches for reaching redistribution", stressing in the same time the "freedom of cooperation and the need for effectiveness in international cooperation". In their view, shared by us, the 1996 declaration came at the right time, in order to spell out that "no distributive or systemic conflict in the field of space applications and cooperation is alive".

7.4 Arguments for the Common Heritage of Mankind

As shown *supra*, law and politics are intertwined. Since the CHM concept lies on the left of the political spectrum, it is favoured by those who support, with various degrees:

- collectivism rather than individualism;
- internationalism instead of nationalism;
- cooperation more than competition;
- the economic interests of the have-nots before those of the haves;
- economic equality more than economic liberty;
- redistribution of wealth instead of acceptance of inequalities;

- equity as opposed to efficiency, or fair outcome as opposed to fair process;
- interventionist versus laissez-faire economics;
- bigger versus smaller government.

All these common traits of socialist ideology can be found in the CHM paradigm.

Many arguments are emotional, making appeal to the "brotherhood of Man". For instance, Indian scholar V.S. Mani (1996, p. 36) hopes that the spacefaring states will "rise above their short term, narrow, Shylockian view of profits" and show instead "magnanimity, compassion and camaraderie to the whole international community". Alan Marshall (1995, p. 52) views the CHM paradigm and its adoption into a regulatory regime as a repellent of space imperialistic tendencies. He believes that the "mildness" of its prescriptions would entail the support of "many, maybe most, space enthusiasts". Those holding a contrary view are seen as either more interested in the personal profit they could make from space endeavours, or eager to forgo the "rights and concerns of many of the world's people" for the sake of outer space expansion. On a similar line of reasoning, Maurice Andem (1999, pp. 7–8) believes that the unpopularity and non-acceptance of the Moon Agreement is due to the fact that individualistic national interest is given priority over the common interest of the humankind as a whole. He considers that the CHM principle conveys a universal truth shared by the main religions, expressing the common origin of the human race.

Francis Lyall (1998, pp. 132–134) agrees that the CHM principle is a major reason behind the lack of a stronger support for the Moon Agreement, and finds some weight in the argument that the have-nots are freeloading. He reckons that the said paradigm is an endeavour by the nations whose technical or other ineptitudes preclude them from exploiting the space resources, to induce spacefaring nations into investing "time, trouble and finance in a project, and then divert to non-participants and non-investors some of the rewards of these entrepreneurial activities". Yet Lyall is wary to fully dismiss the CHM, whose core he views as a moral imperative to preserve and foster if one is to avoid barbarity: "Have' and 'have-not' are too intertwined nowadays for the one to neglect the interests of the other". He is in expectative as to whether the CHM benefits in the law of the sea will indeed reach the peoples of the developing world or will instead be hijacked by their rulers, yet the opportunity for abuse is not considered by Lyall as enough of a reason to shelve the CHM concept. Daniel Goedhuis (1981, p. 8) considers that, while the CHM concept in its space law embodiment did not reduce the inequalities between the developing and developed states to the extent initially sought, it would be nonetheless "difficult to think of an alternative system which, in the light of the present factual situation, could have been devised".

While socialist attitudes from Western analysts are not surprising, what is paradoxical is the eventual lack of support for the Moon Agreement from the very nations it was to benefit. It is also surprising that neither the USSR, nor China or any other Communist nation acceded to it. By 2006, the document has been ratified only by Australia, Austria, Belgium, Chile, Kazakhstan, Lebanon, Mexico, Morocco, the Netherlands, Pakistan, Peru, Philippines, and Uruguay.

As a parenthesis to the CHM paradigm, space collectivism is favoured also at a smaller scale, to some extent on practical grounds. Kalani Chapman (1995) envisages a settlement on Mars modelled on the kibbutz system, where the sense of community will be emphasized "to an infinitely greater extent than the dominant western capitalistic culture on Earth does today'". Chapman's settlement would see the basics for survival (air, water, food and shelter) unconditionally provided for each member. The chief values of the community will be cooperation and mutual aid as opposed to capitalist exploitation. In the early stages – considers Chapman – settlers will work for the sake and the goals of the community, carrying out the functions they excel most at or they most desire: "[E]ach member will contribute what he/she/it can, and each will be provided for ('From each according to his ability, to each according to his needs')". Nonetheless, once the extraterrestrial settlement is largely self-sustaining and automatic systems are established, "people will be free to do as they please".

7.5 Arguments Against the Common Heritage of Mankind

In 1987, US President Ronald Reagan (quoted by Pryor ,2003, p. 23) famously differentiated between communists – those who read Marx and Lenin – and anti-communists – those who understand Marx and Lenin. The critics of the CHM paradigm seem to have a thorough understanding of the flaws plaguing this system.

Addressing the Scottish Unionist Conference in 1948, Winston Churchill (2003, p. 446) called socialism "a deadly fallacy ... the philosophy of failure, the creed of ignorance and the gospel of envy". A criticism of the CHM paradigm is a criticism of communism and socialism. The main argument in opposing the CHM is that communism has failed on earth, and there is no reason this will not happen in outer space. According to Reynolds (1995, p. 118), the anti-market and statist ideas embodied in the Moon Agreement and NIEO have been discredited by the countries that have embraced them. The Soviet Union has ceased to exist, and the former socialist countries of Eastern Europe reverted to democracy and the market economy almost two decades ago. African-American analyst Thomas Sowell (1999, p. 104) wrote that academic Marxists in the US are nonetheless "utterly undaunted" by the collapse of communism throughout Eastern Europe: "Socialism in general has a record of failure so blatant that only an intellectual could ignore or evade it". The arguments *infra* – true as they be – can be equally ignored or evaded.

A number of authors oppose the CHM approach as an antithesis to property rights, seen as essential for the development of space. Michele Hamilton (1988) considers the problems plaguing the paradigm as being obvious; the lack of a strong protection of private property rights eliminates the incentive for private investors to mine the extraterrestrial ore, to develop the know-how associated with this and to get involved in countless other uses of the celestial bodies. She believes that the telecommunications industry would not have reached its current level if the communal approach paradigm would have been applied to satellite development.

Michel Smirnoff (1972, p. 175) criticized a draft of the Moon Agreement, deeming that one needs to be realist when it comes to the terrestrial exportation of the lunar resources:

> If the States which spent the huge sums for the exploration of the Moon will not have the legal possibility to exploit those resources, then they will be confronted with two possibilities: 1 – to stop the explorations or 2 – to continue those explorations ignoring the non-realistic formulations of the Treaties.

According to Alan Marshall (1995, p. 51), a supporter of the CHM paradigm, the prospect that companies could have to rent the site of extraction from the global community would "provoke yelps of horror from space capitalists whose ideological tradition would make them reply that those who take the risk and invest the capital should reap the rewards". From an opposite ideological stand, Listner (2005) considers, indeed, that "[i]t's no wonder that private enterprise is loath to invest in technologies to exploit space with these conditions hanging over their heads". Wassenbergh (1991, p. 82) deems that the CHM concept is too idealistic, as it takes for granted a very strong international solidarity. Whereas such a situation is utopian, each country tending for its own interests, the much-hated capitalist ideological tradition is quite functional.

Another disparagement of the CHM model pertains to the capacity of the developing world to evolve through its own means, and to the progress of the economy. P.J. O'Rourke (1996, p. 154) explains that collectivism does not work due to the faulty economic premise it is based on. In his view:

> There is no such thing as a person's 'fair share' of wealth. The gross national product is not a pizza that must be carefully divided because if I get too many slices, you have to eat the box. The economy is expandable and, in any practical sense, limitless . . . The lesson of economic development is that what happens when we run out of a resource is what happened when we ran out of whale oil – nothing.

A critic of the Communist Manifesto, W.J. Rayment (2000) considers that, while Marx was right in viewing the world in terms of class struggle, he failed to understand that efficiency would lead to an increasing affordability of the means of production. Free societies would elevate the working class to the entrepreneur class, instead of bringing society down to the lowest common denominator. Indeed, Rayment's logic is supported by the evolution of India and the People's Republic of China into spacefaring powers, a tangible proof that developing nations can – and will – participate in the development of space resources. Reynolds (1995, p. 118) remarked as well that space capability is no longer a "superpower monopoly", exemplifying with the credible space programs of the two nations mentioned above, opinion leaders of the developing world. At his turn, space entrepreneur Jim Benson (1998, pp. 48–49) finds the argument that less developed countries are unable to participate in space as currently unsubstantiated. While in the past it was believed that deep space science and resource exploration missions needed to cost billions of dollars, rapidly advancing miniaturization decreased the cost with about an order of magnitude per mission generation, putting it within the reach of any country in the world that wishes to participate at 12 million or less for a deep space experiment.

Indeed, due to the limitless resources of the outer space and to the evolution of the developing world into players on the space arena, it does not matter whether the latter do not hold the pole position in the new commercial space race.

Another reason for criticizing the CHM regime is that it is an expression of the "culture of entitlement". This has been described by Judith Bardwick (quoted in Anderson, 2002), as being:

> an attitude where people believe they do not have to earn what they get. They believe they deserve ... [and] are owed it because of who they are, not because of what they do. In such a culture people take what they have for granted, keep asking for more, and are never satisfied.

Dave Anderson (2002) laments the whole society having moved toward a culture of entitlement over the years, an ethos whose motto is "[w]eaken the strong to strengthen the weak". Keith Urbahn (2005) laments as well the "culture that promotes widespread dependence on government handouts". In his view, a society of entitlement rates justice "not by how much the government encourages those who succeed, but by how much it rewards those who don't". In his view, this equates with a "license for mediocrity", denying individual responsibility and creating convenient excuses for failure. Instead, as explained by Locke (1690, Chapter 5, Section 34), the world was given:

> to the use of the industrious and rational, ... not to the fancy or covetousness of the quarrelsome and contentious. He that had as good left for his improvement, as was already taken up, needed not complain, ought not to meddle with what was already improved by another's labour: if he did, it is plain he desired the benefit of another's pains, which he had no right to, and not the ground ... whereof there was ... more than he knew what to do with, or his industry could reach to.

What is more worrying is that the CHM concept goes even beyond Marxism. In terms of international space economics, the developing countries are not the equivalent of the proletariat – a productive class – but of the lumpenproletariat. The have-nots are not lunar wage-workers whose surplus labour would be exploited by the lunar bourgeoisie. In fact, the true working class of the space economy would be the spacefaring states, upon which the freeloading have-nots would depend as the lumpenproletariat depends upon the bourgeoisie.

7.6 Conclusion

The analysis above, and the terrestrial history of communism, prove that Marxism is a fallacy. Outer space needs to be spared the painful experience of the former Eastern Block. Despite the noble ideals of equity and care for the have-nots, the CHM paradigm has more faults than merits. A refutation of the Common Heritage principle does not mean, however, that the developing world will, or should, be left behind in the space era. China, India and Brazil are living proofs that a developing country can, through its own effort, join the spacefaring club. Instead of freeloading on the efforts of the older spacefarers, the have-nots should pool their meagre financial resources into a common space agency or into regional ones, and proceed at exploiting the riches of outer space for themselves. The rallying cry of Marxism – "Proletarians of all countries, unite; you have nothing to lose but your chains" should evolve into "Countries of the world unite – you have nothing to lose but the chains of gravity". The skies are open.

Chapter 8
Property Status of Extraterrestrial Samples and Extracted Resources

[W]hen the slave says: – the sea is certainly common to all persons – the fisherman agrees; but when the slave adds: – then what is found in the common sea is common property –, he rightly objects, saying: – But what my net and hooks have taken, is absolutely my own –

Hugo Grotius (1608)

8.1 Introduction

The answer to the question "Who owns the Moon" would be incomplete without considering the legal status of material extracted from the Moon. It is to be remarked *ab initio* that the legal treatment applied to real estate and substances removed thereon is different. As Sweet (1882, p. 529) says, "[w]hile unsevered, minerals form part of the land, and as such are real estate. When severed, they become personal chattels".

Whereas the immovables examined *supra* were subject to the *lex situs* of outer space, the current chapter will deal with movables, subject mainly to the *lex domicilii* of the person who caused their removal. However, this is not as simple as it appears. Is the conversion of immovables into movables by way of extraction, allowed in the light of the non-appropriation principle of the Outer Space Treaty? If so, does full ownership of extraterrestrial products vest in those who caused their removal?

Addressing the above questions is not a farfetched intellectual experiment, but a practical matter, in the light of the lunar samples already extracted four decades ago by the *Apollo* and *Luna* missions. While extraterrestrial material often reaches Earth by natural means as meteorites, the Moon is the only celestial body beyond Earth that has been systematically sampled. Between 1969 and 1972, six manned US missions – *Apollo 11, 12, 14, 15* and *17*- brought back 381.7 km of lunar rocks, core samples, sand and dust, originating in six different exploration sites. In addition, three Soviet robotic missions – *Luna 16* (1970), *Luna 20* (1972) and *Luna 24* (1976) – returned 321g of core samples from three other lunar sites (Heiken et al., 1991, pp. 5–7). Three decades would pass until NASA's *Genesis* spacecraft would return samples of solar wind in 2004, and the *Stardust* mission belonging to the same agency would return specks of dust from Comet Wild 2's

V. Pop, *Who Owns the Moon?*, Space Regulations Library Series 4,
DOI 10.1007/978-1-4020-9135-3_8, © Springer Science+Business Media B.V. 2009

coma in 2006. Whereas in November 2005 a Japanese probe, *Hayabusa*, collected materials from near-Earth asteroid 25143 Itokawa, these will be returned to Earth in 2010.

These concrete examples are supplemented by the possible large scale exploitation of extraterrestrial resources and their use either *in situ* or on earth. The new space exploration vision for NASA announced on January 14th, 2004 by the US President urges the "use of lunar and other space resources to support sustained human space exploration to Mars and other destinations" (Aldridge, 2004, p. 52).

8.2 Appropriation of Extraterrestrial Material Under the Outer Space Treaty

Article II of the Outer Space Treaty outlaws appropriation of outer space and celestial bodies "by claims of sovereignty, by means of use or occupation, or by any other means". While the relationship between national appropriation and civil law appropriation of the immovable category of landed extensions has already been discussed *supra*, it is necessary to examine whether the non appropriation principle concerns only the celestial bodies in the territorial (landed extensions) sense, or it applies also to the movable category of extracted resources that originate there. Is the appropriation of extraterrestrial materials by way of extraction an appropriation of the celestial body?

The OST is silent as to the permissibility of appropriating natural materials in the course of exploration and use of the celestial bodies. In the absence of a specific norm clarifying the ownership of lunar resources and samples, scholars held contradictory opinions during the years surrounding the return of the *Apollo* specimens. Several schools of thought can be identified among the space law specialists, some even less liberal than the future Moon Agreement and some very permissive.

8.2.1 First Viewpoint: Outer Space Treaty Prohibits Appropriation of Extraterrestrial Material

The first approach held that several provisions of the Outer Space Treaty would preclude the appropriation of extraterrestrial resources. Given that this view is no longer of currency, it is fitting to talk about it at the past tense.

A number of authors considered that the non-appropriation principle of Article II OST would apply as well to the natural resources of the celestial bodies. Gabriel Lafferranderie (1970, p. 3) deemed that the Outer Space Treaty does not distinguish between celestial bodies and their resources, enouncing only one legal regime prohibiting national appropriation. Gal (1966, p. 47) and Williams (1987, p. 147) cite Stephen Gorove and A. Cocca as other representatives of the view. Indeed, Gorove (1977, p. 247) wrote that the said prohibition may apply to the stock resources of the celestial bodies (as opposed to flow resources) and to exhaustible resources (as opposed to inexhaustible resources). Nonetheless, he considered that the Moon Agreement, in making a distinction between the lunar surface/subsurface and the resources thereon, may have solved the "puzzle" of Article II of the OST, albeit it fails to define the dept of the subsurface and the manner in which the exploitation of the resources embedded in the subsurface can be done "without the exercise of dominion and control over the surface and sub-surface" (Gorove, 1974, p. 30). He also deemed that the contemporary discussion in the UNCOPUOS of a possible moratorium on the exploitation of lunar resources validates the suggestion that the negotiating States consider the ban on appropriation as not applicable to the celestial natural resources (Gorove, 1977, p. 249).

Another tenet brought forth by some scholars in support of the above view is Article I of the OST, providing for the exploration and use of the celestial realms to be carried out in the interest and for the benefit of all countries, as well as Article V of the same treaty deeming the astronauts as envoys of mankind. For instance, Rene Mankiewicz (1968, pp. 164–165) considered that anything taken from the Moon has to be used exclusively in the interest and benefit of all humankind, not for the one of a single State. He agreed nonetheless that taking away the extracted lunar material from a particular State will be very difficult, especially because that State will maintain that their use of such material coincides with the interest of mankind – an explanation he deems dubious. In the year of the Apollo 11 mission, Florence Rusconi (1969, pp. 185–188) asked herself whether the returned samples belong to the USA particularly or to the humankind, and found it "obvious to say that the American astronauts set foot on the moon as envoys from the mankind not as citizens of a powerful country" according to Article V OST, hence all natural stuff found on the Moon and other celestial bodies will be considered as part of mankind's patrimony. Her argument is however contradicted by the real purpose of Article V of the Outer Space Treaty that is the provision of assistance to astronauts in distress regardless of their nationality. A different interpretation would forbid astronauts, for example, to go to orbit in order to repair a private satellite. Gabriel Lafferranderie (1970, p. 6) wrote one year later that the State that utilises and holds extraterrestrial materials is the depositary of this patrimony that is indivisible.

8.2.2 Second Viewpoint: Outer Space Treaty Allows Appropriation of Limited Quantities of Extraterrestrial Material

Some authors are more reserved, deeming that, while the non-appropriation principle applies to lunar materials, a limited use in quantity and scope would be exceptionally permitted. In the same time, an extended use of the extraterrestrial materials would amount to national appropriation.

Eugene Brooks (1968, pp. 346–347), for instance, considered that several legal consequences can be logically drawn from the Outer Space Treaty – such as the susceptibility of consumption for local needs of all local resources required for the operation of a station, and the permissibility for a State to use the amount necessary for such support. In his view, the use and carrying away of extraterrestrial substances should be limited to that necessary for the purpose of gathering of scientific information and knowledge, and such knowledge should be shared, pursuant to Article IX of the OST, with the UN General Secretary, the international scientific community, and the public. One year later, he wrote that the conflict between the freedom of use and the non-appropriation principle is "only potential" and does not become of actuality until the use extraterrestrial resources has reached a certain threshold or they have been used beyond the minimal needs of scientific testing or *in-situ* mission support. Brooks was confident that no such critical point has been reached through the removal of a limited quantity of lunar material by the Apollo 11 mission He nonetheless deemed the private prospecting of minerals as constituting illegal appropriation in the view of the common benefit clause of Article I OST (Brooks, 1969, pp. 160–169).

For Oscar Fernandez-Brital (1969, p. 197), the possibility of extracting a large quantity of materials from the Moon presented legal difficulties. He believed that such transport of extraterrestrial materials is allowed by the OST without claims, yet only in small quantities for scientific research. "Total consumption" – deemed he – must never occur, as it would run counter Articles II and IX of the OST. Marko Markoff (1972, p. 169) viewed the exchange of limited quantities of lunar samples, with scientific aims, between the US and USSR as an exception lying beyond the scope of the normal applicability of Article II of the OST. Another exception was seen as being the utilization in "reasonable quantities" of extraterrestrial materials for the support of a mission.

Although Vladimir Kopal (1972, p. 154) found it obvious that the activity of collecting samples and removing them from the Moon serves the common interest of all mankind in the spirit of the Outer Space Treaty, he feared that this practice, if expanded too much, might be considered as a violation of the non-appropriation principle.

8.2.3 Third Viewpoint: Outer Space Treaty Allows Appropriation of Extraterrestrial Material

Most authors offer more generous opinions as to the property rights in extraterrestrial resources and samples, drawing a clear distinction between the appropriation of outer space and celestial bodies, and the appropriation of materials thereon. To them, the seizure of natural resources merely forms part of the freedom of exploration and use of outer space, by analogy with the rules underlying the freedom of the high seas (Williams, 1987, p. 147). In the present context, the rights of private enterprises to explore and exploit the natural resources of the moon, without having *fee simple* ownership over the land thereof, could be generically called "enterprise rights". Article I of the OST consecrates the freedom of scientific investigation, exploration and use of the extraterrestrial realms, whereas the "residuary rule of presumptive freedom of action" (Lauterpacht, 1975, p. 220) proclaims that what is not prohibited is permitted.

Indeed, a strong opponent of national appropriation of the high seas, Grotius (1608) did not discourage the appropriation of the resources found there. In support of this view, he found it appropriate to cite the play *Rudens* ("The Rope") by Titus Maccius Plautus:

> [W]hen the slave says: – the sea is certainly common to all persons – the fisherman agrees; but when the slave adds: – then what is found in the common sea is common property – he rightly objects, saying: – But what my net and hooks have taken, is absolutely my own.

By their public domain character, celestial bodies are similar with the seashore under Roman law, common and accessible to all by the law of the nature (Justinian's Institutes, Book II, Title I, para. 1.) While, as a *res communis*, the seashore was incapable of ownership (Thomas, 1975, p. 75), not the same regime was valid for the "pebbles, precious stones and the like which are found on the seashore" ; these "at once become the finder's property by the law of nations" (Justinian's Institutes Book II Title I, para. 18). By analogy, the moon rocks "found" on the moon would become the finder's property.

As shown above, the act of extraction transforms the minerals – part of the land while unsevered – into "personal chattels" (Sweet, 1882, p. 529) or movables, and pass as such under the national jurisdiction of the extractor. Fasan (1962, p. 11) very clearly states that "[t]he mined up mineral (*fructus separatus*) will become property of the exploiter, because it will have been separated from the soil of the celestial body itself". The Hungarian doctrine is like-minded; Laszlo Szaloky (1969, p. 178) deems that the act of separation and development endows the natural resources with a "separate legal destiny", allowing them to be "taken into ownership by occupation". The materials and energies found in the extraterrestrial realms may be used freely for providing for man's stay there, for building purposes and may be "freely exploited". The extractor, in Szaloky's view, may process them for transportation and may pile them up for that purpose. Marta Miklody (1979, pp. 179–180) finds "unrealistic" to refute a country the right of possessing the discovered substances according to its laws, as the non – appropriation principle concerns the "exclusion of sovereignty and not the civil law of possession". Gyula Gal (1969, p. 200) deems that, in the light of the *res omnium communis* character, the return of extraterrestrial samples and their appropriation by the extracting state can be done without acquiring sovereignty over the Moon: "[e]xploitation of the fish of the high seas and the minerals of the sea-bottom rests on the same legal ground" ; he is constant in his views, stating three decades later (1996, p. 47) that the "*res communis omnium* character" imposed by Article II OST upon the celestial bodies prohibiting the acquisition of property does not apply to the objects thereon.

In Argentina, Manuel Ferrer (1969, pp. 145–146) is of the strong opinion that the OST, while banning sovereignty claims, has never – nor could have legitimately – forbidden the seizure by States of "pieces of celestial bodies" or "portions separated therefrom", these being susceptible of appropriation in accordance with their nature of res communis. Such appropriation, believes he, is necessary for carrying out the exploitation of a celestial body. One year earlier, Ferrer (1968, p. 146) deemed that the State that has established a lunar station may use the local resources for its building and maintenance, this right being obvious from the legitimacy of establishing such a station in the first place.

Stephen Gorove's (1985, p. 227) views evolved from the ones shown *supra*, agreeing that there is "little doubt" in the view of the position adopted by the UN-COPUOS that the non-appropriation principle of the OST refers to an area and not to the resources found there. C. Wilfred Jenks (1969, pp. 148–149) believes too that the non-appropriation principle applies to the celestial realms as such, and not to their resources, on whose respect it "may be necessary or desirable to recognize that title thereto can be acquired by extraction or use". Writing immediately after the Apollo 11 mission, he considered that neither the 1963 Declaration of Legal Principles, nor the OST answer in a clear and direct manner to the question of ownership of the samples retrieved by Armstrong and Aldrin. He nonetheless concluded that, pursuant to the freedom of use of the extraterrestrial realms embodied in Article I, "the answer must surely be that the ownership vests in the Government of the United States through the action of its agents in reducing into possession moon samples", and that, "[s]o far as is yet known the United States assumes that it has property" in the samples. The same author considers that Article I directs the United States to "facilitate and encourage international co-operation" in scientific investigation, and that the US appears to be acting upon this principle, intending to make the extracted samples available to scientific bodies throughout the world "for purpose of scientific investigation without any transfer of property".

The municipal law of the United States explicitly qualifies the *Apollo* samples as property. Thus, a NASA policy directive states that "[i]t is . . . NASA policy to . . . retain, as property of NASA, all lunar samples allocated for public display purposes" (NPD 1387.2F, para. 1.g); another directive mentions "Federal and NASA regulations pertaining to the safeguarding of these materials as Government property" (JPD 7100.1F, Section 1). As owner of the samples, the United States Government is free to dispose of them; *abusum*, as the ultimate attribute of ownership involving the right to destroy the object of property rights, is exercised by NASA when providing samples for destructive analysis (Gooding and Connolly, 1995) involving their consumption, contamination or modification. In the same time, *jus abutendi* is exercised by the US Government when relinquishing ownership; as early as 1969, the US President declared that "moon samples symbolical of the common interest of mankind in the Moon will be presented to other states" (Jenks, 1969, p. 149).

The Agreement on Space Cooperation between NASA and the Soviet Academy of Sciences signed at Moscow on January the 21st, 1971 provided, *inter alia*, for the possibility to conduct a coordinated exchange of lunar samples. Accordingly, USA and USSR exchanged on several occasions in 1971 and 1972, various samples collected by the *Apollo* and *Luna* missions (Markoff, 1972, p. 167). According to Vassilevskaya (1977, p. 474), the Soviet Union used to regularly provide a certain quantity of lunar samples to different states for tests at no cost, this being done in accordance with the spirit of the OST. Most important, Russia has commercialized lunar material; in December 1993, Sotheby's auctioned three small particles of regolith retrieved by one of the Soviet probes. The fragments, weighing about one carat (200 mg), sold for US$442,500 – i.e. US$2.2 million per g (Arthur, 1998).

Commenting on the exchange of samples between USSR and USA, Marko Markoff (1972, p. 169) erroneously considered that this does not imply the appropriation of the planetary matter returned to earth, the State proceeding to the exchange being merely a holder of lunar materials. Nonetheless, as *nemo dat quod non habet*, USSR and USA could not have alienated the samples were they not the owners. The sale on the market of a *Luna* sample has also set a precedent of commercialization of material that has been extracted from a celestial body. Lunar material is, legally, *res in commercio*. As no objections were been voiced by third States, it can be stated that, as an attribute of ownership, the right to commercialize extraterrestrial material has entered into customary international law. Indeed, Gyula Gal (1996, p. 47) finds worthy of mentioning "some facts of international practice", namely that both USA and USSR collected and returned lunar samples – objects that "were appropriated by U.S. and Soviet authorities respectively and have been owned by them without objection from the international community."

All of the lunar samples returned to Earth by the *Apollo* missions have been collected by twelve astronauts. One may ask whether the act of extracting them could have vested ownership in them. An answer may be found in John Locke's Second Treatise of Civil Government, where:

> by virtue thereof [of the original law of nature], what fish any one catches in the ocean ... is, by the labour that removes it out of that common state nature left it in, made his property who takes the pains about it (Locke, 1690, Section 30).

Following Lockean logic, did the labor performed in removing these from the state of nature vest ownership in the twelve labourers that physically extracted them? Where they the only ones who "took the pains about" the removal of the lunar samples? Let us read further, in Section 43:

> For 'tis not barely the ploughman's pains, the reaper's and thresher's toil, and the baker's sweat, is to be counted into the bread we eat; the labour of those who broke the oxen, who dug and wrought the iron and stones, who felled and framed the timber employed about the . . . utensils, which are a vast number, requisite to this corn, from its being seed ... to its being made bread, must all be charged on the account of labour, and received as an effect of that.

From the above, it can be drawn that the twelve astronauts do not and cannot own the samples they removed. Behind their labour invested in the collection of each sample stand the enormous physical and intellectual effort of 400,000 people directly involved in building the *Apollo* Moon ships (Chaikin, 1994, p. 11), and the financial effort of the American taxpayers that have spent $25.4 billion dollars towards this goal. Jim Gooding and John Connolly (1995) quantify the work that went into acquiring these samples:

> 6 Saturn V rockets; 1,104,500 gallons of kerosene propellant; 2,363,000 gallons of liquid oxygen; 1,707,400 gallons of liquid hydrogen; 36,260 gallons of hypergolic propellants; 3 lunar rovers; 6 command module pilots; 12 moonwalkers; and a large supporting cast.

One of the issues discussed by the space law scholars, as seen *supra*, is whether Article V of the OST precludes the private or national appropriation of extraterrestrial material. Whereas, as provided by the said norm, States Parties to the OST have to regard the astronauts as "envoys of mankind in outer space", H. Nauges (1979, p. 270) asks whether an envoy of mankind can carry out work which would profit only a single State. The quantified labour *supra* is joined to the unquantified one of the whole humankind. As US President Richard Nixon (1973) symbolically recognizes and symbolically rewards with lunar samples:

> as we stretch for the stars, we know that we stand also upon the shoulders of many men of many nations . . . In the deepest sense our exploration of the moon was truly an international effort.

According to Roby (1902, p. 456), Roman Law provided that one can acquire possession through free persons if they are one's agents and act with the intention of acquiring it for their master. This happens if an agent has a mandate; otherwise, the master acquires the possession only on ratification. In a valid construction, authors argue that the astronauts act as agents for their government, that is the owner of their labor and the one entitled to what the astronaut's labour is joined to. To Jenks (1969, pp. 149–151), the ownership of the *Apollo* samples vests in the US Government "through the action of its agents in reducing into possession moon samples" and, as long as space expeditions consist of "persons in the service of and acting for their governments", he deems that "any property which they may acquire in lunar minerals by extraction will presumably vest, by law, service regulation, or contract, in the government employing them" . However, "[i]f any question were to arise of moon adventurers acquiring personally or for any corporate body a right of property in what they brought back from the moon, the range of problems relating to the nature and status of such property would become more complex".

8.3 Appropriation of Extraterrestrial Material Under the Moon Agreement

All the material returned from the Moon by USA and USSR has been collected after the entry into force of the Outer Space Treaty, and prior to the entry into force of the Moon Agreement. As neither country is a signatory to the latter, the *lex lata* status of lunar material extracted is to be analysed in the light of the Outer Space Treaty.

The Moon Agreement is, however, not completely irrelevant. Some of its provisions restate in fact the customary law developed following the *Luna* and *Apollo* missions. On the other hand, the Moon Agreement may become the legal standard in the event that one of its States Parties returns extraterrestrial material, although no sample activities have been performed so far by any of them. In the same time, given the lesser liberalism of the Moon Agreement as a whole, it results *a fortiori* that what is allowed by the Moon Agreement must be permitted also by the Outer Space Treaty, even if not expressly.

Whether they agree or not with the CHM language of the MA, a number of authors welcomed some of its provisions and some interpreted them in a very permissive way. In analysing the use of extended amounts of lunar material for the construction of solar power satellites, Rene Oosterlinck (1996, p. 276) considers that the *in place* formula of the Moon Treaty means that "even for mining activities of this scope i.e. some 600,000 Tons per year this would not be considered as an appropriation of part of the Moon". In Peter Haanappel's (1980, p. 30) view, the right to collect samples was not universally recognized under the Outer Space Treaty regime, while the Moon Agreement clarifies the status of samples removed from the Moon. Rene Oosterlinck (1996, p. 276) equally considers the relevant Moon Treaty provisions as a step forward in the ruling of natural resources of outer space, as they authorise explicitly the removal from the Moon of samples. Stanley B. Rosenfield (1980, p. 71) sees the Moon Agreement as "the first outer space treaty specifically granting the right to collect and remove and use substances from outer space." However, he doubts that this would augment the right of exploration and use already granted by Article I of the OST, and he states that the practice to that date has established the custom of collecting and removing of samples. Vladimir Kopal (1972, p. 157), more correctly, sees the Moon Agreement as bringing "a significant innovation concerning samples which is based on the present practice".

8.3.1 Appropriation of Extraterrestrial Material for Scientific Investigation

Article 6.2 of the Moon Agreement gives States Parties, "[i]n carrying out scientific investigations and in furtherance of the provisions of this Agreement", the right to "collect on and remove from the moon samples of its mineral and other substances". Such samples may be used, according to the same tenet, "for scientific purposes", and "shall remain at the disposal of those States Parties which caused them to be collected". The States Parties are also urged to "have regard to the desirability of making a portion of such samples available to other interested States Parties and the international scientific community for scientific investigation".

The main aim of extraterrestrial sample return is, at least for now, scientific investigation. The scientific benefits derived from the lunar samples are invaluable as, while meteorites are chance samples of solar system debris, the lunar samples can be keyed to known locations on the Moon (Heiken et al., 1991, pp. 5–6). All lunar materials returned have provided invaluable clues in the scientific knowledge of our natural satellite, ranging from understanding planetary formation dynamics to key refinements in in-situ resource utilization technology theory (Elliott, 1998).

This norm restates in fact the customary law developed following the *Luna* and *Apollo* missions. While the text does not clarify whether "making available" implies transfer of ownership or just a loan, state practice has provided for both situations.

The above paragraph does not clarify either whether this is to be done free of charge or for a sum of money. State practice is more generous that the Moon Agreement also in so far as it has involved donations of samples both for scientific purposes, as provided in the Moon Agreement, and for symbolic purposes.

While the state practice was evolving towards an unequivocal answer, codification by means of treaty was going towards the same direction. In 1974, during the Moon Agreement negotiations in the UNCOPUOS Legal Subcommittee, a proposal by Egypt, India and Nigeria to vest in the United Nations property rights to the Moon samples was rejected (Thomas, 1980, p. 160).

Carl Christol (in United States Senate, 1980, p. 203) comments that, "[w]ithout the use of such words as 'property' and 'ownership', [the norm] did not interpose inhibitions upon the exclusive use by the space-resource States of the identified substances". To Silvia Williams (1987, p. 147), the above norm "implies a certain right of property on the part of the States concerned even though the Agreement carefully avoids using such terms". Similarly, to Gyula Gal (1996, p. 45) it "carefully avoids such terms for objects collected and removed as property, ownership, though above stipulation grants certain rights of an owner to the states concerned". Peter Haanappel (1980, p. 31) comments that the Moon Agreement "remains silent on the question of ownership of such samples removed from the Moon" ; however, he reads the "at the disposal" provision together with the prohibition of appropriation of resources "in place" by Article 11.3, concluding that a sample, "once removed, becomes the property of the collecting State, which is under the obligation to use it in accordance with the terms of Article VI(2)."

The approach used by the Moon Agreement for scientific samples is functional. A sample is not defined according to its dimensions, but according to its scientific usage. As to the difference between scientific research and commercial use, David R. Criswell (in United States Senate, 1980, p. 228) considers there is a "blurring of scientific usage in the 'lunar sample research' meaning and in the industrial meaning", as scientific usage might comprise the use of extraterrestrial materials to develop new commercial processes. Carl Christol (in United States Senate, 1980, p. 203) believes that, over time, the extent and nature of scientific investigations may be "far-ranging", hence the above norm "will allow for very substantial uses of natural resources."

The second principle of the norm – namely the call for sharing samples with the scientific community in other countries – is seen by Okolie (1980, p. 63) as a means of reinforcing the equality principle and embodying the concerns of the developing world regarding the shareable uses of extraterrestrial resources, while to the L-5 society, the provision is "merely hortatory" insofar as "creates no specific enforceable obligation; it differs little in effect from the provisions of Article IX of the Outer Space Treaty establishing a principle of 'co-operation and mutual assistance' for outer space activities" (United States Senate, 1980, p. 126.). Indeed, the above norm is very soft insofar as it does not impose a legal obligation to share scientific samples with other States. The obligation is of a moral nature (Oosterlinck, 1996, p. 276; Gal, 1996, p. 47), States being not expected to share samples, but merely to consider the possibility of doing so. Marko (1992–1993, p. 334) suggests replacing this with an obligation of providing samples "upon the request of a State Party, and if reasonably available", this language requiring sharing of materials but leaving open the possibility that unreasonable requests need not be honoured.

8.3.2 Appropriation of Extraterrestrial Material for Support of Scientific Missions

Article 6.2 of the Moon Agreement allows States Parties to use:

> in the course of scientific investigations ... mineral and other substances of the Moon in quantities appropriate for the support of their missions.

The appropriation of these resources is qualified, being subject both to qualitative and quantitative restrictions. The approach used by the Moon Agreement for *in-situ* resources utilization (ISRU) is therefore both functional ("in the course of scientific investigations ... for the support of their missions") and quantitative ("in quantities appropriate").

The language of the MA implies that only missions pursuing scientific investigations may be supported by extraterrestrial materials. In the light of the above norm, appropriation of extraterrestrial resources would be permitted for the building of scientific stations and vehicles, for their supply with fuel, and for sustaining life thereupon. Should a hotel be established on a celestial body, it would not qualify, as space tourism is not scientific investigation. Nonetheless, it would be absurd to expect commercial users to import building materials and supplies from earth while freely available locally for lunar scientists.

In the same time, not any quantities may be appropriated, but only that which is necessary for the support of the mission; no surplus would be permitted. Commentators are more generous, nonetheless; Damodar Wadegaonkar (1984, p. 34) leaves the quantitative assessment at the sole appreciation of the States conducting the mission, while expecting that important quantities of extraterrestrial material may be taken away and not mere samples. He deems that the norm – an express affirmation of the freedom of scientific investigation – does not define how long the mission may last or the kind of mission envisaged. According to Stephen E. Doyle (1998, p. 119), the extent of the use of the in-situ resources is "determinable by the mission definition or mission profile, defining its purposes and objectives". Charles Okolie (1980, p. 63) interprets "quantities appropriate" as implying "the right to recover investment made by interested parties in Moon exploration within the range of return on investment deemed as 'appropriate'", the whole permission for *in-situ* resource utilization being seen as a norm which "protects public and private investors in space exploration". Rene Oosterlinck (1996, p. 276) sees this norm as one of the steps forward in the ruling of natural resources of outer space, notwithstanding the shortfalls of the MA.

The US official position on the provision, as expressed by Delegate Hosenball to the UNCOPUOS on July 3rd 1979 (quoted in Rao, 1981, p. 276), is that "there is no limit upon the rights of States parties of those natural resources found on celestial bodies as are appropriate for the support of their mission", and that this may well promote the likelihood of the commercial or practical exploitation of the natural resources of the celestial bodies.

8.3.3 Appropriation of Extraterrestrial Material for Commercial Purposes

Article 11.3 of the Moon Agreement prohibits that, on the Moon, "natural resources in place shall become property of any State, international intergovernmental or non-governmental organization, national organization or non-governmental entity or of any natural person". A *per a contrario* reasoning demonstrates that, whereas no claims can be asserted to the natural resources that make up with the lunar territory one whole, private ownership of resources once no longer in place (mined/extracted) is permitted (Dekanozov, 1980, p. 5; Rothblatt, 1983, p. 243; Haanappel, 1980, p. 30). Rene Oosterlinck (1996, p. 276) explains that the "in place" language has been successfully proposed by the United States in the course of the Moon Agreement negotiations.

Such appropriation is nonetheless limited by the Common Heritage of Mankind provisions of the same document. As analysed in a separate chapter, Article 11.7.d of the MA provides for "[a]n equitable sharing by all States Parties in the benefits derived from [the natural resources of the Moon]", whereas Article 11.5 urges States Parties to establish a regime to govern the exploitation of lunar resources "as such exploitation is about to become feasible". The language of Article 11.5 has been subject to academic scrutiny as to its prohibiting or not the exploitation of the lunar resources prior to the establishment of a regime to govern such activity, or in the event such a regime would not be established. A moratorium for the exploitation of the extraterrestrial resources would place them in the legal category of *res extra commercium* pending the creation of the said regime that would turn them into *res in commercio*.

According to Peter Haanappel (1980, p. 30), the appropriation of extracted materials may happen both prior to the international regime, and after its establishment – only that in this latter case the appropriation would be qualified by the terms of the regime. Milton L. Smith (1988, p. 47) believes that the moratorium issue would have never surfaced unless the exploitation of lunar resources was permitted by the OST. In the opinion of K. Narayana Rao (1981, p. 278), it is irrelevant whether a moratorium clause is present or not in the MA given that the Moon and the resources thereon are already vested in the humankind, while failure to establish an international regime will not divest them from the said owner. In his view:

> [t]he only consequence of such a failure will be that the realisation of the Moon's resources by mankind remains in suspended animation. In any event the States Parties are definitely prohibited from reaching the Moon for exploiting its resources.

Most of the authors consider there is no such temporary prohibition. Karl-Heinz Bockstiegel (1981, p. 214) deems that neither Article 11 of the Moon Agreement, nor the document as a whole, give any grounds to conclude that there is a moratorium. He believes that, were it impossible to agree, for political reasons, on the said regime for an extended period of time, this would not preclude any private enterprise or State from using space resources as soon as economically feasible. Hanneke van Traa-Engelman (1980, p. 76) believes that there is no "explicit prohibition" to exploit the lunar resources before the establishment of an international regime, yet the CHM provisions combined with "the expressed impossibility to obtain property rights outside an international regime" will most certainly inhibit exploitation of the extraterrestrial resources by the States Parties or the private enterprises thereon.

In April 1979, the American representative to the legal subcommittee of the UNCOPUOS declared (according to Rao, 1981, p. 76) that his country was not ready to agree to an "express or implied" prohibition on exploiting the lunar natural resources prior to the international conference meeting and agreeing on the "appropriate machinery". Three months later, on July 16th, delegate Neil Hosenball addressed the UNCOPUOS on behalf of the US, stating that the draft Moon Agreement does not place any moratorium upon the exploitation of the lunar resources pending the establishment of an international regime, allowing thus "orderly attempts" to ascertain, by means of experimental beginnings followed by pilot operations whether the exploitation of the lunar resources is practical and feasible (UN Doc A/AC.105/PV.203, quoted by Finch, 1979, p. 123). In Peter Haanappel's view (1980, pp. 30–32), the fact that the above statement was not contradicted in the consensus-based UNCOPUOS means that it received the "blessings" of the other delegations. His own view is that the language of Article 11 "makes clear" that there is no moratorium; "rather" – says he – "the right of exploitation is recognized" . This did not stop US Senators Frank Church and Jacob K. Javitts to contact in October 1979 the US Secretary of State Cyrus Vance and to voice their fears that the MA would result in a "de facto moratorium" on space resources activities. In his reply, the Secretary of State remarked that the concept of a moratorium has been opposed both by the USSR and the US during the meetings of the UNCOPUOS. Goedhuis also noted that the UNCOPUOS itself rejected such a moratorium, and that the MA is silent about it (Goedhuis, 1981, pp. 2–8).

Most important, as remarked by Haanappel (1980, p. 30) and accepted by Rao (1981, p. 275), the developing countries did not push for a moratorium; were they eager to establish such a temporary prohibition, they would have done it explicitly as they did in the case of the deep seabed resources. Indeed, the academic discussion of the existence of a lunar moratorium may have been sparked by a document having actually imposed such a measure for the deep seabed. On December 15th 1969, the UN General Assembly adopted Resolution 2574 (XXIV), whereby it declared that, pending the establishment of an international regime including appropriate international machinery governing the exploitation of the resources of the area:

(a) States and persons, physical or juridical, are bound to refrain from all activities of exploitation of the resources of the area of the sea-bed and ocean floor, and the subsoil thereof, beyond the limits of national jurisdiction;
(b) No claim to any part of that area or its resources shall be recognized.

As shown *supra*, the CHM regime as envisaged for the Moon is not the same with the CHM regime as per the UNCLOS in its 1982 and 1994 incarnations. In 1978, Sri Lankan diplomat Christopher W. Pinto (quoted in Mom, 2005, pp. 19–20) explained that the minerals subject to the [sea] CHM regime cannot be freely mined:

> They are not there, so to speak, for the taking. The common heritage of mankind is [the] common property of mankind. The commonness of the 'common heritage' is a commonness of ownership and benefit. The minerals are owned in common by your country and mine, and by all the rest as well. In their original locations these resources belong in undivided and indivisible share to your country and to mine, and to all the rest – to all mankind, in fact, whether organized as a state or not. If you touch the nodules at the bottom of the sea, you touch my property. If you take them away means to take away my property.

Indeed, pursuant to the 1969 moratorium resolution, the minerals recovered from the deep seabed could not be appropriated pending the establishment of a regime, and then, according to Article 137.2 of the UNCLOS, only in accordance with Part XI of the Convention and the "rules, regulations and procedures of the Authority". Almost three centuries before being formulated, the above arguments have been rejected by Locke (1690, Chapter 5, Section 29):

> By making an explicit consent of every commoner necessary to any one's appropriating to himself any part of what is given in common, children or servants could not cut the meat which their father or master had provided for them in common without assigning to every one his peculiar part. Though the water running in the fountain be every one's, yet who can doubt but that in the pitcher is his only who drew it out? His labour hath taken it out of the hands of nature, where it was common, and belonged equally to all her children, and hath thereby appropriated it to himself.

L.F.E. Goldie (1976, p. 289) uses the same Lockean logic:

> Arguments to the effect that the resources of outer space, like those of the deep seabed, are the 'common heritage of the mankind' does not necessarily prohibit their exploitation or appropriation. The right of access of all to a common property resource is well known. A villager who draws a bucket of water from the common well regards that water as his own in exactly the same sense that a seabed miner, or a future miner of outer space may regard the minerals he obtains.

8.4 The Property Regime of Meteorites

In its Article 1.3, the Moon Agreement explicitly excludes from its material jurisdiction the "extraterrestrial materials which reach the surface of the earth by natural means". This provision refers to meteorites, natural objects that fall from space and survive to land on Earth's surface. More than 10,000 meteorites have been collected, greatly surpassing the quantity of extraterrestrial samples returned by space missions. Being mostly of asteroidal provenience, a number of meteorites of lunar and Martian origin have been found, and speculations have been advanced regarding meteorites originated on Mercury and Venus. While still in space, the Sun-orbiting pieces that are potential meteors and meteorites are called meteoroids; should these burn up in the atmosphere without surviving to reach the ground, they would be classified as meteors (Lewis, 1996, pp. 87–90).

According to Nathan Goldman (1985, p. 231, note 26), the Moon Agreement "clearly legitimates the appropriation of meteorite[s], celestial bodies which have struck the Earth". Vladimir Kopal (1972, p. 155) deems the provision as being "undisputable". We view the above clause as simply excluding the meteorites from the *ratione materiae* jurisdiction of the Moon Agreement. While outside the regulation of international space law, ownership of meteorites pertains to the domain of municipal law. Krystyna Wiewiorowska (1980, p. 85) agrees that "there is a general trend to accept that meteorites should be recognized as belonging to the State on the territory on which they have fallen". By ratifying the Moon Agreement, States would not have to modify the existing domestic regimes concerning meteorites; in the same time, the Moon Treaty would not have any impact over an eventual treaty regarding the international regime of meteorites.

While meteorites are outside the material jurisdiction of the international space law, the municipal law regarding ownership of meteorites is far from being homogenous. Several contrasting regimes have crystallized, reflecting the lack of uniformity in approaching property rights throughout the globe.

The first stance considers meteorites as parts of the realty, vesting the property of a meteorite in **the landowner on whose property it fell**. This is the case in the United Kingdom, where, according to Philip Bagnall (1992, p. 45), one can "sell it, donate it to the Natural History Museum or keep it as an unusual paperweight". In the United States, judicial decisions favour as well the landowner (Martinez, 1998). In *Goddard v. winchell (1892)*, the court held that a meteoric event equates to accretion on a planetary scale, hence a fallen meteorite becomes the property of the person on whose land it fell by virtue of becoming part of his soil. As a note, accretion is in fact a slow process, whereas avulsion is a sudden one. A similar decision attributing property of the meteorite to the land owner at the expense of the finder was delivered in *Oregon Iron Co. v. Hughes* (1905).

The second view, in contrast, considers meteorites as movable *res nullius*, vesting the property of a meteorite in **its finder**. In Japan, the civil code awards ownership of the meteorites to the finder (Schmitt, 2001). The same is the *de facto* regime in Antarctica. Wiewiorowska (1980, p. 85) knows of "no indication which shows the legal regulation of meteorites fallen outside the States' jurisdiction". Indeed, the Antarctic Treaty System, for instance, does not contain any provisions regarding meteorites found at the antipodes. Nevertheless, the collection of Antarctic meteorites is likely to qualify as "scientific investigation", free under Article II of the 1959 Antarctic Treaty. This activity would not be hindered therefore by the 1991 Protocol on Environmental Protection to the Antarctic Treaty, as scientific research activities are not "Antarctic mineral resource activities", according to Article 1.7. of the 1998 Convention on the Regulation of Antarctic Mineral Resource Activities. The practice regarding the 11 lunar meteorites totalling about 2 kilograms collected on the Antarctic ice cap by expeditions of the USA and Japan (neither of these countries having territorial claims in the Antarctic) between 1979 and 1991 (Heiken et al., 1991, pp. 6–9) would indicate that they belong to the finding nation.

Both Japan and the USA are nonetheless under the legal obligation of Article III.1.c of the Antarctic Treaty to exchange and make freely available, to the greatest extent feasible and practicable, scientific observations and results from Antarctica – hence data on the Antarctic meteorites.

A third stance sees meteorites as public goods, vesting their property in **the state on whose territory they fell**, regardless of who owns the land on which it fell or who discovered it. This is not unusual, given that several domestic regimes vest ownership of certain minerals in the State. Sweet (1882, p. 529) talks about "Royal mines" under English law, i.e. mines of gold or silver, who "belong to the Crown, in whosoever land they may be found". Meteorites fallen on the territory of Denmark – with the exception of Greenland – belong to the state, the finder receiving in exchange a reward based on the weight, condition and rarity of the meteorite (Bagnall, 1992, p. 45). Meteorites are provincial property in the Chaco Province in Argentina, by virtue of the provincial Constitution. In India, meteorites belong by law to the Geological Survey, without any compensation being paid to the landowner or the finder, whereas in Switzerland a meteorite is state property but its finder is paid a compensation "not higher than the object's value" (Schmitt, 2001). This regime is justified by the scientific importance of the meteorites and, as long as appropriation by the state entails compensation for the finder or the landowner, it is a fair practice under the doctrine of eminent domain.

A *lex ferenda* proposal advanced by Argentine scholar Enrique Ferreyra (quoted in Williams, 1969, p. 180) would vest property of the meteorites in **the United Nations**. Whereas the scientific interest of the international community in all meteorites is justified, not many arguments can be brought sustaining actual ownership of meteorites by the United Nations. We consider that the best legal regime would have a high regard both for the scientific importance of the extraterrestrial material and for the property rights of the landowner and the finder. An eminent domain appropriation by the State doubled by payment of a fair compensation to both the landowner and the finder may well work to the satisfaction of all parties.

8.5 Conclusion

The conversion of immovables into movables by way of extraction is allowed in the light of the non-appropriation principle of the Outer Space Treaty, a distinction being made between the appropriation of outer space and celestial bodies, and the appropriation of materials thereon. While lacking *fee simple* ownership over the land thereof, private actors are, under the Outer Space Treaty regime, entitled to "enterprise rights" – that is, to explore and exploit the natural resources of the moon. Ownership of extraterrestrial products vests in those who sponsored their removal, through the labor invested in seizing them. As an attribute of ownership, the right to commercialize extraterrestrial material has entered into customary international law. The municipal law of the United States explicitly qualifies the *Apollo* samples as property, whereas material extracted by Soviet probes has lawfully entered the free market.

Whereas the Moon Agreement is far from being generally accepted, it does codify some of the customary law developed after the *Apollo* and *Luna* missions. Given its lesser liberalism as a whole, it results a fortiori that what is allowed by the said document must be permitted also by the Outer Space Treaty, even if not expressly. The Moon Agreement clarifies the material/territorial distinction in the matter of non-appropriation, allowing ownership of extracted resources, the collection of scientific samples and that of materials for in-situ support of missions. The document imposes no moratorium on the exploitation of lunar resources pending the establishment of an international regime.

While outside the material jurisdiction of the international space law, the municipal law regarding the ownership of meteorites is far from being homogenous. The existence of contrasting regimes reflects both the lack of uniformity in approaching property rights throughout the globe, and the difficulty in finding a universally acceptable solution to the issue of extraterrestrial sample and mineral resources ownership.

Chapter 9
Conclusion

Having started this volume with the tale of the Little Prince, it is fitting to finish it off with another saga – the one of the wise men of Chelm. In Yiddish folklore, the Chelmites are known for their unconventional wisdom – read foolishness. One of the many tales narrates about their love for the moon and their frustration at seeing it waxing and then waning. Desiring to capture it for themselves, the endeared simpletons cover a well where the moon had been glimpsed, only to find it next day back in the sky.

Like the "wise men of Chelm", many gullible customers would buy corner lots on the moon, only to find them on sale, years later, by another swindler. Whereas in terms of space law the purported sale of extraterrestrial real estate is a trivial issue, we deemed necessary to critically address and dismantle with legal arguments this issue, given that the popularity of the Lunar Embassy and similar ventures has hijacked the public perception of the law of outer space. As detailed *supra*, the logic behind these ventures is flawed; if a State may not appropriate the extraterrestrial realms, this prohibition is extended *a fortiori* to all of its subjects. Furthermore, we have shown that a mere claim is not tantamount with ownership – both *animus* and *corpus possidendi* are necessary in order to call a good one's own. The host of subsequent unsubstantiated claims by countless wannabe landlords is yet another argument in dismantling the issue of "unreal estate". Neither Dennis Hope, nor his predecessors or copycats, own the moon.

Besides calling the consumer protection agencies to address this matter, we have offered that the rise of the "Lunar Embassy" highlights the need for a clarification of the legal status of the extraterrestrial realms. Why does the law seem to be silent when it comes to property rights outside the atmosphere of the earth? We found necessary to scrutinize whether the factual situation on the extraterrestrial realms calls for legal regulations. We examined the sources of law in their dual dimension – that is, the facts that have caused and shaped the law of extraterrestrial real estate, and the norms which express this law – finding that the norms and rules regarding property rights in the celestial realms are rather limited and fail to define basic concepts such as celestial body. We have also found out that, although dismissed by a part of the academia, politics are crucial in shaping the law.

The Outer Space Treaty contains in Article II a fundamental principle, outlawing the national appropriation by any means of outer space and celestial bodies.

V. Pop, *Who Owns the Moon?*, Space Regulations Library Series 4,
DOI 10.1007/978-1-4020-9135-3_9, © Springer Science+Business Media B.V. 2009

We deemed necessary to find out what is the precise meaning of "celestial bodies", pondering whether asteroids and comets are immovable land-like territorial extensions that cannot be legally appropriated, or floating movable goods, capable of being captured and reduced into private ownership. We considered the employment of the spatialist and functionalist approaches, the use of the criterion of actual movability from orbit by human action, and original theories such as the analogy between the legal status of asteroids and icebergs, and we offered that some extraterrestrial resources are not, legally speaking, celestial bodies – hence their appropriation ought to be permitted.

In order to find an answer to the question "who owns the moon", we deemed necessary to enquire what "owning" means. We explained that property is not a monolithic construct, but a bundle of rights or of attributes (*jus utendi*, *jus fruendi* and *jus abutendi*), its essence boiling down to the control over access – in other words, to the right of excluding others from using a good. We then proceeded to examine whether the non-appropriation principle on the international plane, as contained in the Outer Space Treaty, results also in the prohibition of the appropriation of the celestial bodies on the plane of private property rights. We offered that, regardless whether prohibited or not, landed private property cannot survive outside the sphere of sovereignty. What can exist is a mere factual situation that will or will not be subsequently converted into property by means of State endorsement – which is prohibited *de lege lata.*

In the first chapter, we have shown that Dennis Hope does not own the moon. But who actually does? We answered with the statement that, under the current incarnation of space law, "everybody and nobody" owns the moon. Far from being Orwellian doublethink – that is, simultaneously accepting two mutually contradictory beliefs –, the fact that the moon is owned by "everybody and nobody" hails from the "bundle of rights" theory of property.

De lege lata, the extraterrestrial realms pertain to the legal category of *res communis*, whereby use is permitted ("everybody owns it") yet title is denied ("nobody owns it"). While private appropriation of the extraterrestrial realms may not exist on its fullness under a *res communis* regime, some of its elements are permitted. Between planetary structures and the soil of the celestial bodies there is a layer of "legal insulation" – no physical interaction between the two can change the legal status of the other. The Outer Space Treaty establishes among the States Parties an open access and free use regime on the Moon, making it a public good where *jus utendi* is not absolute, having to be enjoyed with due regard to the corresponding interests of other users and likely users. In a nutshell, *res communis* implies freedom for an actor to use a good; as long as one uses this good, one may not be obstructed by another, yet this latter is free to use the good as soon as the first has ceased to use it. In the extraterrestrial realms, the regime is a hybrid of *res communis* – at international level – and *res publica* at municipal level, given the need for a nationally issued license for private actors. Whereas the Outer Space Treaty establishes several rules, most of these remain at the stage of principle. Many open access regimes are self-regulatory, and where the text of the law is silent, custom is bound to develop.

We have pointed out that the *status quo* of the commons regime is not very stable, being challenged on two fronts. On the left, the adepts of the Common Heritage of Mankind (CHM) paradigm have succeeded in causing the United Nations General Assembly to pass the Moon Agreement, effectively planting the Marxist standard in the lunar soil. This egalitarian regime, which not only prohibits private landed property but also calls for an "equitable" sharing of *jus fruendi* overseen by a managerial entity, offers no incentive to exploit the extraterrestrial ores when the investors have to share the benefits with free riders who believe in a culture of entitlement. Together with John Locke, we believe that law ought to provide for the "industrious and rational" and not for the "quarrelsome and contentious". The arguments brought forth in this book prove that the CHM paradigm has more faults than merits, and that outer space needs to be spared the painful experience of the ex-communist nations. As a practical solution to the needs of the developing countries, we suggested *supra* that they pool their resources into a common space agency or into regional ones, and proceed at exploiting the riches of outer space for themselves.

We have then seen how, from the other end of the political spectrum, scholars and activists maintain that property rights – comprising *jus abutendi* – are a useful engine and, in all likelihood, a precondition for pushing forward the development of the extraterrestrial realms. In the tradition of the American Frontier, whose settlement was encouraged by governmental plans of privatizing the public domain, the proponents of the private property of space seek a similar privatization of the extraterrestrial realms. We have shown that the frontier spirit can translate in a dual way in the settlement of space. The first approach is that of "crossing the Alleghenies" and establishing an anarcho-capitalist way of life on the Moon, based on individualism – where, as with the Preemption Act and the Spitzbergen Treaty, the possession can be later offered the protection of a sovereign entity. The second approach is that of an authority privatizing the commons, in the manner of the United States Homestead Act of 1862. Together with John Locke, we see such privatization – provided there is still enough left for the others – as an active administration of the public trust stemming from the common heritage. Whereas the CHM paradigm is *lex lata* for those who ratified the Moon Agreement, the frontier paradigm is, for now, *lex ferenda*, at least in the United States where the Aldridge Commission called for guaranteeing appropriate property rights for the entities seeking to develop space resources.

The frontier paradigm has proven its worth on our planet, and it most likely will do so in the extraterrestrial realms. Securing property rights would be more beneficial to humankind, compared to the alternative of keeping the extraterrestrial realms undeveloped. In the same time, support for property rights in outer space needs to be complemented by support for property rights on earth, given the ongoing erosion of the fee simple ownership on our planet.

The answer to the question "Who owns the Moon" would be incomplete without considering the legal status of material extracted from the Moon. Although fee simple ownership is prohibited, under the *res communis* regime the conversion of immovables into movables by way of extraction is allowed in the light of the non-appropriation principle, a distinction being made between the appropriation of outer space and celestial bodies and the appropriation of materials thereon. While lacking *fee simple* ownership over the land thereof, private actors are, under the Outer Space Treaty regime, entitled to "enterprise rights" – that is, to explore and exploit the natural resources of the moon. Ownership of extraterrestrial products vests in those who sponsored their removal, through the labor invested in seizing them. As an attribute of ownership, the right to commercialize extraterrestrial material has entered into customary international law. Whereas the Moon Agreement is far from being generally accepted, it does clarify the material/territorial distinction in the matter of non-appropriation, allowing ownership of extracted resources, the collection of scientific samples and that of materials for *in-situ* support of missions. The document imposes no moratorium on the exploitation of lunar resources pending the establishment of an international regime.

We have been warned at the beginning of this work that space law demands an Olympian level of wisdom, and that attempts at addressing its challenges are not easier than hacking down a mountain with a blunt knife. We have also been warned that property is a metaphysical creature, a phantom figure that dissolves into the void as soon as one tries to grasp it – much like the fabled moon of Chelm.

It must be that the Sphinx took pity on Oedipus, the furies were compassionate with Sisyphus, and the Gods cared for Heracles; none of them had to address the legal status of Selene's silver chariot. Nor were the literates of ancient Greece entrusted with this task.

But, if they were, what would have they spoken of? Most certainly, Aristophanes would have laughed at the comedy of the Lunar Embassy. He would have shown us how ridiculous it is to sell Selene's horses just because one thought so and wanted so. Sophocles, at his turn, would have anticipated Aristotle's tragedy of the commons, and would have told us how that which is common to all has the least care bestowed upon it. We would have laughed, and we would have wept. Yet we would not have been inspired. We would have needed a Homer to tell us about heroes climbing Olympus, about Archimedes moving the heavens with a fulcrum stretching to infinity, about people settling a new frontier with the assurance nobody will take away the fruit of their labour and the fields they toil.

We would have realized, after paying careful attention to the three literates, that the choice between a space comedy, a space tragedy and a space odyssey, is obvious.

Bibliography

Acland, J. (chairman), 2003. *Walking Access in the New Zealand Outdoors*. A report by the Land Access Ministerial Reference Group, Ministry of Agriculture and Forestry, Wellington, New Zealand.

Adam, D., 2005. Property rights on the Moon. *The Guardian*, 24 October.

Agence France-Presse, 2005. Russian astrologer cleared to sue NASA over plans to bombard "sentimental" comet. *Spacedaily* [online], 6 May. Available from: http://www.spacedaily.com/2005/050506134341.nt1q1zco.html [accessed 17 September 2006].

Ahuja, A., 2000. Who owns the Moon? *The Times*, *Times 2*, 14 September, p. 13.

A-Infos News Service, 2002. Anarchist Question and Answer. *Anarchist Age Weekly Review* [online]. No. 491, 25 March. Available from: http://www.ainfos.ca/02/mar/ainfos00625.html [accessed 17 September 2006].

Aldridge, E.C. (chairman), 2004. *A Journey to Inspire, Innovate, and Discover: Report of the President's Commission on Implementation of United States Space Exploration Policy*. US Government Printing Office.

Allard, M.-P., 2001. *The Retroactive Effect of Conditional Obligations in Tax Law* [online]. Available from: http://www.bijurilex.org/site/export/concord/e/allard_2001/allard_2001.txt1.htm [accessed 17 September 2006].

Andem, M.N., 1999. Twentieth anniversary of the 1979 Moon Treaty: the legal status of the Moon and other celestial bodies revisited in the light of the commercialization of outer space activities, *Proceedings of the 42nd Colloquium on the Law of Outer Space*, p. 383.

Anderson, D., 2002. Changing a culture of entitlement into a culture of merit. *The CPA Journal* [online], November. Available from: http://www.nysscpa.org/cpajournal/2002/1102/nv/nv8.htm [accessed 17 September 2006].

Annan, K., 1997. *Renewing the United Nations: A Programme for Reform*. Report of the Secretary-General. 14 July (Document A/51/).

Annan, K., 2005. "In larger freedom": decision time at the UN. *Foreign Affairs*, Vol. 84, Issue 3, p. 63.

Applied Space Resources, 1999. *Policy Statement* [online]. Available from: http://appliedspace.com/property_rights.htm [accessed 13 January 2000].

Aristotle, 350 BC. *Politics*.

Arthur, C., 1998. Jewellery's final frontier. *The Independent*, 11 May.

Arutunyan, A., 2005. Russian astrologist plans to crash NASA's Independence Day. *MosNews* [online], 19 April. Available from: http://www.mosnews.com/feature/2005/04/19/deepimpact.shtml [accessed 17 September 2006].

Associated Press, 1944. Do you want the Moon? Why not file claim with Interior Dept.? *The Lowell Sun*, 14 December, p. 14.

Associated Press, 1946. Tells how to homestead on the Moon – (maybe). *The Nebraska State Journal*, 26 January.

Associated Press, 1948. Two claim Moon – bids are rejected. *Pottstown Mercury*, 1 April, p. 1.

Associated Press, 1954. Arkansas trio hopes to sell lots on Mars. *The Lima News*, 8 September, p. 3.

Associated Press, 1955. Cold eye cast on selling of Moon land. *The Post Standard*, 23 November, p. 2.

Associated Press, 1962. Believes soviets could claim Moon. *Stevens Point Daily Journal*, 16 March, p. 11.

Associated Press, 1969. Moon lots for sale. *The Chronicle Telegram*, 18 August, p. 21.

Associated Press, 1975. Man charged in selling 1,000 acre Moon lots. *The Lincoln Star*, 24 May, p. 8.

Associated Press, 1998. NASA apologises to Navajos for plan to put remains on Moon. *The Salt Lake Tribune,* 14 January.

Baca, K.A., 1993. Property rights in outer space. *Journal of Air Law and Commerce*, Vol. 58, p. 1041.

Bagnall, P., 1992. Psst! You wanna buy a piece of moon rock? *New Scientist*, 20 June, p. 45.

Bayross, M., 2005. *Исковое заявление. №2 (после 4 июля - космического террора НАСА) о запрещении псевдонаучной деятельности НАСА в РФ, создающей опасность причинения вреда в будущем и компененсации морального вреда.* Available from: http://www.marinabai. ru/preobrajenie2.files/page0005.htm [accessed 5 January 2006].

Benkö, M., Schrogl, K.-U., 1997. Space law at UNISPACE III (1999) and beyond. *Proceedings of the 40th Colloquium on the Law of Outer Space*, p. 157.

Benson, J.W., 1998. Space resources: first come first served. *Proceedings of the 41st Colloquium on the Law of Outer Space*, p. 46.

Bentham, J., 1802. *The theory of legislation.*

Betancourt, C.H.R., 1997. The Treaty of 67 in front of 21st century, *Proceedings of the 40th Colloquium on the Law of Outer Space*, p. 309.

Bilder, R.B., Cameron, E.N., Kulcinski, G.R., Smith, H.J.H., 1989. Legal regimes for the mining of helium-3 from the Moon. *Wisconsin Centre for Space Automation and Robotics*, Madison, WI, WCSAR-TR-AR3-8901-1.

Blum, Y.Z., 1965. *Historic Titles in International Law*. The Hague: Nijhoff.

Bockstiegel, K.-H., 1981. Energy from space – chances and limitations from a legal point of view. *In: Earth-Oriented Space Activities and Their Legal Implications, Proceedings of the Symposium held on October 15–16, 1981*. Montreal: McGill University, p. 204.

Bray, P.M., 1999. *Public Trust Doctrine (PTD)* [online]. Available from: http://www.braypapers.com/PTD.html [accessed 17 September 2006].

Brooks, E., 1966. National control of natural planetary bodies: preliminary considerations. *Journal of Air Law and Commerce*, Vol. 32, p. 315.

Brooks, E., 1968. Control and use of planetary resources. *Proceedings of the 11th Colloquium on the Law of Outer Space*, p. 339.

Brooks, E., 1969. Legal aspects of the lunar landing. *Proceedings of the 12th Colloquium on the Law of Outer Space*, p. 160.

Brown, G., 1999. *Environmental Law and Natural Resource Policy Law – Course Syllabus.* Alaska Pacific University, AK.

Bueckling, A., 1979. The strategy of semantics and the "Mankind provisions" of the Space Treaty. *Journal of Space Law*, Vol. 7, p. 15.

Burke, E., 1775. *Speech of Edmund Burke, Esq., on Moving his Resolutions for Conciliation with the Colonies* [online], March 22. Available from: http://www.econlib.org/LIBRARY/LFBooks/Burke/brkSWv1c3.html [accessed 17 September 2006].

Burn, E.H., 1994. *Cheshire and Burn's Modern Law of Real Property*. 15th ed. London: Butterworths.

Buyuranus.com, 2000. *About Buy Uranus* [online]. Available from: http://www. buyuranus.com/about.php4 [accessed 17 September 2006].

Card, O.S., 1995. *Earthfall*, Tor.

Carson, D., 1965. Advice for Moon homesteaders. *The Edwardsville Intelligencer*, 23 August, p. 3.

Central Press Canadian, 1955. Selling acre lots on the Moon. *The Lethbridge Herald*, 2 December, p. 15.

Chaikin, A., 1994. *A Man on the Moon*. London: Penguin.

Chapman, K., 1995. Thoughts from a settlement on Mars. *The Manoa Journal of Fried and Half-Fried Ideas (About the Future)*, Vol. 6 Available from: http://www.soc.hawaii.edu/future/j6/Mars.html.

Cheng, B., 1957. International law and high altitude flights: balloons, rockets and man-made satellites. *International and Comparative Law Quarterly*, Vol. 6, p. 487.

Christol, C.Q., 1997. The Moon Treaty and the allocation of resources. *Annals of Air and Space Law*, Vol. XXII-II, p. 31.

Churchill, W., 2003. *Never give In! – The Best of Winston Churchill's Speeches*. New York: Hyperion.

Clark, S., 2000. Stars in his eyes. *New Scientist*, 14 October.

Clarke, A.C., Kubrick, S., 1999. *Foreword to 2001: A Space Odyssey*. London: Orbit.

CNN, 1997. 3 Yemenis Sue NASA for Trespassing on Mars. *CNN* [online], July 24. Available from: http://www.cnn.com/TECH/9707/24/yemen.mars/ [accessed 17 September 2006].

Cocca, A.A., 1958. Principles for a declaration with reference to the legal nature of the Moon. *Proceedings of the 1st Colloquium on the Law of Outer Space*, p. 34.

Cocca, A.A., 1962. Basic statute for the Moon and heavenly bodies. *Proceedings of the 5th Colloquium on the Law of Outer Space*, p. 5.

Cocca, A.A., 1986. The common heritage of mankind: doctrine and principle of space law – an overview. *Proceedings of the 29th Colloquium on the Law of Outer Space*, p. 17.

Cocca, A.A., 1995. Reservation of a lunar zone for SETI purposes. *Proceedings of the 38th Colloquium on the Law of Outer Space*, p. 270.

Coffman, R.P., 1938. Uncle Ray's corner: a little Saturday talk. *Appleton Post Crescent*, 18 June, p. 12.

Coke, E., 1628. *The First Part of the Institutes of the Laws of England, or, a Commentary on Littleton*. London

Cole, L., 1996. The men who claimed the Moon. *Mimosa*, Vol. 18, p. 32.

Collins, L. (Ed.), 1987. *Dicey and Morris on the Conflict of Laws*. 11th ed. London: Stevens & Sons.

Commission on Global Governance, 1994. *Our Global Neighbourhood: The Report of the Commission on Global Governance*.

Cramer, K., 1997. The lunar users union – an organisation to grant land use rights on the Moon in accordance with the Outer Space Treaty. *Proceedings of the 40th Colloquium on the Law of Outer Space*, p. 352.

Csabafi, I., 1962. Current problems of space law in 1962. *Proceedings of the 5th Colloquium on the Law of Outer Space*.

Csabafi, I., 1971. *The Concept of State Jurisdiction in International Space Law*. The Hague: Nijhoff.

CSDC, 1992. LUNA 2010 (aka Option 51). *California Space Development Council Meeting*, 17 February.

Danilenko, G.M., 1998. The concept of the "Common Heritage of Mankind" in international law. *Annals of Air and Space Law*, Vol. XIII, p. 247.

Davis, K.C., 2005. *The Anti 1967 UN Space Treaty Page*, [online]. Available from: http://kevincdavis.net/AntiUn/why.htm [accessed 17 September 2006].

De Saint-Exupéry, A., 1946. *Le Petit Prince*. Gallimard.

Deepsea Ventures, Inc., 1974. Notice of discovery and claim of exclusive mining rights, and request for diplomatic protection and protection of investment. 14 November. *International Legal Materials* (1975), Vol. 14, p. 51.

Dekanozov, R.V., 1974. Juridical nature of outer space, including the Moon and other celestial bodies. *Proceedings of the 17th Colloquium on the Law of Outer Space*, p. 200.

Dekanozov, R.V., 1977. Draft treaty relating to the Moon and the legal status of its natural resources. *Proceedings of the 20th Colloquium on the Law of Outer Space*, p. 197.

Dekanozov, R.V., 1980. Juridical nature and status of the resources of the Moon and other celestial bodies. *Proceedings of the 23rd Colloquium on the Law of Outer Space*, p. 5.

Dembling, P., 1997. Negotiating issues in forming the 1967 Treaty on outer space. *Proceedings of the 40th Colloquium on the Law of Outer Space*, p. 34.

Diederkis-Verschoor, I.H.Ph., 1980. Some more observations on solar satellites and their legal aspects. *Proceedings of the 23rd Colloquium on the Law of Outer Space*, p. 149.

Dietrich, G.B., Goldstein, W.C., 1998. Collective trusteeship for near space: the case for UNNESA. *Space Policy*, Vol. 14, p. 9.

Dinkin, S., 2004. The Dinkin Commission report (Part 1). *The Space Review* [online], 21 June. Available from: http://www.thespacereview.com/article/164/1 [accessed 17 September 2006].

Dixon, M., 1998. Promise me the Moon – property rights in space. *Accountancy,* Vol. 121, Issue 1254, p. 29.

Doyle, S.E., 1997. Concepts of space law before Sputnik. *Proceedings of the 40th Colloquium on the Law of Outer Space*, p. 3.

Doyle, S.E., 1998. Using extraterrestrial resources under the Moon Agreement of 1979. *Journal of Space Law*, Vol. 26, p. 111.

Dupuy, R.-J., 1983. Conclusions in the light of the notion of Mankind. In: *The Settlement of Disputes on the New Natural Resources*. The Hague: Hague Academy of International Law/United Nations University/Nijhoff.

El Centro de Talca, 2004. TV rescata a talquino dueño de la Luna. *El Centro de Talca*, 29 April.

Elliott, B., 1998. Private commercial lunar sample return missions. *The 17th International Space Development Conference*, 24 May, Milwaukee, Wisconsin.

Embassy of Canada at Washington, 1974. Canadian Response to Deepsea Ventures, Inc. 6 December. *International Legal Materials*, (1975), Vol. 14, p. 67.

Emerson, R.W., 1862. American civilization. *The Atlantic Monthly*. Vol. IX, Issue 54, p. 502, April.

Esquivel De Cocca, M., 1997. Is it necessary to redefine principles and concepts of the Outer Space Treaty? *Proceedings of the 40th Colloquium on the Law of Outer Space*, p. 84.

Fabun, D., 1952. The Little Men and the Moon. *Rhodomagnetic Digest*, Vol. III, Issue 19, p. 3.

Fasan, E., 1962. Law and peace for the celestial bodies. *Proceedings of the 5th Colloquium on the Law of Outer Space*, p. 9.

Fasan, E., 1966. Legal problems for celestial bodies and their solution by the IISL. *Proceedings of the 9th Colloquium on the Law of Outer Space*, p. 54.

Fasan, E., 1980. Some legal problems regarding the Moon. *Proceedings of the 23rd Colloquium on the Law of Outer Space*, p. 9.

Fasan, E., 1984. Large space structures and celestial bodies. *Proceedings of the 27th Colloquium on the Law of Outer Space*, p. 243.

Fasan, E., 1994. Human settlements on planets – new stations or new nations. *Journal of Space Law*, Vol. 22, p. 47.

Feather, C.E., 2004. No fooling: Geneva citizens claimed ownership of the Moon back in April 1966. *Ashtabula Star Beacon*, 1 April.

Fernandez-Brital, O., 1969. Activities on celestial bodies, including exploitation of natural resources. *Proceedings of the 12th Colloquium on the Law of Outer Space*, p. 195.

Ferrer, M.A., 1968. La problematica juridica de una estacion lunar. *Proceedings of the 11th Colloquium on the Law of Outer Space*, p. 145.

Ferrer, M.A., 1969. Introductory report on activities on celestial bodies and the exploitation of natural resources. *Proceedings of the 12th Colloquium on the Law of Outer Space*, p. 141.

Filho, J.M., 1996. Private, State and international public interests in space law. *Space Policy*, Vol. 12, p. 59.

Finch, E.R., 1979. 1979 United Nations Moon Treaty encourages lunar mining & space development. *Proceedings of the 22nd Colloquium on the Law of Outer Space*, p. 123.

First Millennial Foundation, 1997. *Why the First Millennial Foundation*. Rifle, Co: First Millennial Foundation.

Fitzpatrick, J., 1991. Home Reef Fisheries Development: A Report from Torres Strait. *Cultural Survival Quarterly*, Vol. 15, Issue 2, p. 18.

Florida Geographic Alliance, 1992. *Tribal Society: The Masai* [online]. Available from: http://fga.freac.fsu.edu/academy/pdf/tribal_fact_sheet2.pdf [accessed 17 September 2006].

Frankman, M.J., 1993. Fractals and the Common Heritage of Humanity. *The 9th Annual Conference of the Canadian Association for the Study of International Development, June 1993, Carleton University*.

Gabrynowicz, J.I., 2005. The international space treaty regime in the globalization era. *Ad Astra*, Fall 2005, p. 30.

Gal, G., 1966. Les realites techniques et juridiques concernant les corps celestes. *Proceedings of the 9th Colloquium on the Law of Outer Space*, p. 55.

Gal, G., 1969. *Space Law*. Sijthoff/Oceana.

Gal, G., 1996. Acquisition of property in the legal regime of celestial bodies. *Proceedings of the 39th Colloquium on the Law of Outer Space*, p. 45.

Gal, G., 1997. Thirty years of functionalism. *Proceedings of the 40th Colloquium on the Law of Outer Space*, p. 125.

Galactic Mining Industries, 2006. *Space Property Rights and the Future* [online]. Available from: http://www.space-mining.com/2004GU9.html [accessed 4 May 2008].

Genta, G., Rycroft, M., 2003 Genta, G., Rycroft, M., 2003. *Space, the Final Frontier?* Cambridge: Cambridge University Press.

Geon, B.S., 1997. A right to ice?: The application of international and national water laws to the acquisition of iceberg rights. *Michigan Journal of International Law*, Vol. 19, p. 277.

Gernsback, H., 1962. Why go to the Moon. *Family Weekly*, 14 January, p. 4.

Goedhuis, D., 1978. The changing legal regime of air and outer space. *International and Comparative Law Quarterly*, Vol. 27, p. 576.

Goedhuis, D., 1981. Report on the desirability of states ratifying the Moon Treaty of 1979. *São Paulo Conference on the Law of the World, August 16–21, 1981*. Washington, DC: World Peace through Law Center.

Goldie, L.F.E., 1969. Legal fictions in the development of blueprints for law to govern activities in outer space. *Proceedings of the 12th Colloquium on the Law of Outer Space*, p. 153.

Goldie, L.F.E., 1976. Is there a general international law of original ownership? The possible relevance of general doctrines governing the possession of deep ocean-bed resources. *Proceedings of the 19th Colloquium on the Law of Outer Space*, p. 287.

Goldman, N.C., 1985. Space activities: transforming space law. *Proceedings of the 28th Colloquium on the Law of Outer Space*, p. 227.

Golrounia, A.A., Bahrami, M., 1997. Outer Space Treaty in 21st century: a change of concept. *Proceedings of the 40th Colloquium on the Law of Outer Space*, p. 303.

Gooding, J., Connolly, J., 1995. R.S.V.P. for a lunar sample. *Beyond L.E.O*, Vol. 2, no. 3.

Gorbiel, A., 1985. Questions of analogies between sea and space international law. *Proceedings of the 28th Colloquium on the Law of Outer Space*, p. 142.

Gorove, S., 1968. Interpreting article ii of the outer space treaty. *Proceedings of the 11th Colloquium on the Law of Outer Space*, p. 40.

Gorove, S., 1974. Property rights in outer space: focus on the proposed Moon Treaty. *Journal of Space Law*, Vol. 2, p. 27.

Gorove, S., 1977. Sovereign rights in outer space. *Proceedings of the 20th Colloquium on the Law of Outer Space*, p. 244.

Gorove, S., 1983. Major concerns of private enterprise regarding recent developments in space law. *Advances in the Astronautical Sciences*, Vol. 53, p. 199.

Gorove, S., 1985. Private rights and legal interests in the development of international space law. *Space Manufacturing*, Vol. 5, p. 226.

Gorove, S., 1992. Major definitional issues in the space agreements. *Proceedings of the 35th Colloquium on the Law of Outer Space*, p. 76.

Gray, K., 1991. Property in thin air. *Cambridge Law Journal*, Vol. 50, p. 252.

Gray, K., 1993. *Elements of Land Law*. Second edition. London: Butterworths.

Grossman, H.I., 2004. *Property Rights Under Anarchy*. Working paper, Brown University.

Grotius, H., 1608. *The Freedom of the Seas, or the Right which Belongs to the Dutch to Take Part in the East Indian Trade*. New York: Oxford University Press.

Haanappel, P.P.C., 1980. Article XI of the Moon Treaty. *Proceedings of the 23rd Colloquium on the Law of Outer Space*, p. 29.

Hackworth, G.H., 1940. *Digest of International Law*. Vol. I, Washington, DC: US Government Printing Office.

Haeruman, H., 1995. Environmental dimensions of non-wood forest products. In: FAO, *Report of the International Expert Consultation on Non-Wood Forest Products*, Rome.

Hamilton, M., 1998. The application of the Fifth Amendment to property rights in outer space. *NSS Conference 'Property Rights and Commercial Space Development'*, 2 December, Washington.

Hanover, W.H., 1961. Sovereignty in space. *Proceedings of the 4th Colloquium on the Law of Outer Space*, p. 119.

Hara, R., 1983. Legal status of astronauts and other personnel on the Moon. *Proceedings of the 26th Colloquium on the Law of Outer Space*, p. 165.

Hara, R., Stanczyk, J., 1984. Space law and the Roman law concepts. *Proceedings of the 27th Colloquium on the Law of Outer Space*, p. 51.

Hardin, G., 1968. The tragedy of the Commons. *Science*, Vol. 162, Issue 3859, p. 1243.

Hartzok, A., 2000. *Welfare or Earthshare? That is the Question!* [online]. Available from: http://www.earthrights.net/docs/earthshare.html [Accessed September 21, 2006].

Hashimoto, Y., 1993. The status of astronauts towards the second generation of space law. *Proceedings of the 36th Colloquium on the Law of Outer Space*, p. 206.

He, Q., 1997. The Outer Space Treaty in perspective. *Proceedings of the 40th Colloquium on the Law of Outer Space*, p. 51.

Heidmann, J., 1994. What legal questions are raised by the establishment of a dedicated lunar far side specific crater for high sensitivity radioastronomy? *Proceedings of the 37th Colloquium on the Law of Outer Space*, p. 255.

Heiken, G.H., Vaniman, D.T., French, B.M. (Eds.), 1991. *Lunar Sourcebook: A User's Guide to the Moon*. Cambridge: Cambridge University Press.

Heim, B.E., 1990. Exploring the last frontiers for mineral resources: a comparison of international law regarding the deep seabed, outer space, and Antarctica. *Vanderbilt Journal of Transnational Law*, Vol. 23, p. 819.

Hervy, I., 1972. De la Terre a la Lune sans passeport: pour un droit transcedental de l'espace. *Proceedings of the 15th Colloquium on the Law of Outer Space*, p. 145.

Hoffstadt, B.M., 1994. Moving the heavens: lunar mining and the "Common Heritage of Mankind" in the Moon Treaty. *UCLA Law Review*, Vol. 42, p. 575.

Hofmann, M., 2000. Space cemeteries – a challenge for the legal regime of outer space. *Proceedings of the 43rd Colloquium on the Law of Outer Space*.

Hope, D.M., 1997. *You own the what? – A True Life Story and Adventure*. The Lunar Embassy.

Hoppe, H.-H., 2001. *Democracy – The God that Failed: Studies in the Economics and Politics of Monarchy, Democracy, and Natural Order*. New Brunswick: Transaction Publishers.

Hudgins, E.L., 1998. Martian Law. *Cato White Papers and Miscellaneous Reports* [online], 15 August. Available from: http://www.cato.org/pubs/wtpapers/980815paper.html [accessed 17 September 2006].

Humboldt Standard, 1952. Claim rights to Moon minerals. *Humboldt Standard*, 28 February, p. 25.

Husain, Y., 2000. Top ten reasons to inhabit outer space. *Space.com* [online], 21 July. Available from: http://www.space.com/scienceastronomy/solarsystem/reason_02_000723.html [accessed 17 September 2006].

Hyde, D., 2005. Sorites paradox. In: E.N. Zalta (Ed.), *The Stanford Encyclopaedia of Philosophy* (Fall 2005 Edition) [online]. Available from: http://plato.stanford.edu/archives/fall2005/entries/sorites-paradox/ [accessed 17 September 2006].

Inoshita, D., 1957. Japan Astronautical Society enjoying boom business in real estate on Mars. *The Coshocton Tribune*, 16 October, p. 7.

International Institute of Space Law, 2004. *Statement by the Board of Directors of the International Institute of Space Law (IISL) on Claims to Property Rights Regarding the Moon and Other Celestial Bodies*, [online]. IISL. Available from: http://www.iafastro-iisl.com/downloads/Statement_Moon_2004.doc [accessed 17 September 2006].

International News Service, 1957. Tokyo agent reports rush for Mars land. *Mansfield News Journal*, 22 November, p. 1.

Jasentuliyana, N., 1997. The development of the outer space treaties and legal principles from a Third World perspective. *Proceedings of the 40th Colloquium on the Law of Outer Space*, p. 57.

Jenks, C.W., 1958. *The Common Law of Mankind*. London: Stevens & Sons.

Jenks, C.W., 1965. *Space Law*. London: Stevens & Sons.

Jenks, C.W., 1969. Property in Moon samples and things left upon the Moon. *Proceedings of the 12th Colloquium on the Law of Outer Space*, p. 148.

Jessup, P.C., Taubenfeld, H.J., 1959. *Controls for Outer Space and the Antarctic Analogy*. New York: Columbia University Press.

Johnson, E., 1946. In Hollywood. *The Port Arthur News*, 3 November, p. 9.

Kaczmarczyk, A., 2003. *Cyberdemocracy Ante Portas* [online]. Warsaw: Institute of Mathematical Machines. Available from: http://www.imm.org.pl/pdf.php?id=61 [accessed 18 September 2006].

Keller, J., 1959. The challenge of the Moon. *The Newark Advocate*, 31 March.

Knapp, D., 1996. Lunar real estate sale. *CNN San Francisco*, 7 March. Available from: http://www.cnnsf.com/newsvault/output/lunar.html [accessed 28 November 2000].

Kopal, V., 1972. The development of legal arrangements for the peaceful uses of the Moon. *Proceedings of the 15th Colloquium on the Law of Outer Space*, p. 149.

Lachs, M., 1964. The international law of outer space. *Recueil des cours, Hague Academy of International Law*, Vol. 113 (1964-III), p. 1.

Lafferranderie, G., 1970. *Le regime juridique applicable aux materiaux provenant de la lune et des autres corps celestes – rapport introductif*. Groupe de travail sur le droit de l'espace du C.N.R.S.

Lansing, R., 1917. A unique international problem. *American Journal of International Law*, Vol. 11, p. 763.

Lauterpacht, H. (Ed.), 1955. *Oppenheim's International Law*. Vol. I. 8th Ed. London: Longmans Green & Co.

Lauterpacht, H., 1975. *International Law ('the collected papers')*. Vol. 2. Cambridge: Cambridge University Press.

Lay, S.H., Taubenfeld, H.J., 1970. *The Law Relating to Activities of Man in Space*. Chicago: The University of Chicago Press.

Leich, M.N., 1980. *Digest of United States Practice in International Law*. Washington, DC: Department of State.

Lenin, V.I., 1917. *The Tasks of the Proletariat in Our Revolution*. Sankt Petersburg: Priboi Publishers.

Lewis, J.S., 1996. *Mining the Sky: Untold Riches from the Asteroids, Comets, and Planets*. Reading, Mass.: Addison-Wesley.

Listner, M.J., 2005. It's time to rethink international space law. *The Space Review* [online], 31 May. Available from: http://www.thespacereview.com/article/381/1 [accessed 17 September 2006].

Locke, J., 1690. *The Second Treatise of Civil Government*. London.

Lotko, E.A., 1968. *Address Delivered at the 20th Anniversary Party Celebrating Founding of the Nation*. 22 December, Archives of the Nation of Celestial Space.

Lunar Embassy, 2001. *Newsflash from the Lunar Embassy Staff* [online]. March 2001–May 2001. Available from: http://www.lunarembassy.com/ls/lenewsflashpast_e.shtml [accessed 20 March 2001].

Lunar Embassy, 2006a. *The Frequently Asked Questions List* [online]. Available from: http://www.lunarembassy.com/ls/legeneralfaq_e.shtml [accessed 17 September 2006].

Lunar Embassy, 2006b. *Extraterrestrial Property and Space Law: Fact and Fiction* [online]. Available from: http://www.lunarembassy.com/ls/lespacelaw_e.shtml [accessed 17 September 2006].

Lyall, F., 1998. On the Moon. *Journal of Space Law*, Vol. 26, p. 129.

Mackenzie Stuart, A., 1932. *The Law of Trusts*. Edinburgh: W. Green & Son.

Magno, P., 1972. Communication. *Proceedings of the 15th Colloquium on the Law of Outer Space*, p. 65.

Magno, P., Verdacchi, C., 1976. Outer space resources. *Proceedings of the 19th Colloquium on the Law of Outer Space*, p. 320.

Mangan, J.T., 1956. *Report to the Universe: The First Seven Years – A White Paper*. 30 April, State Department, Celestia, USA.

Mangan, J.T., 1958. *State of the Sky December 1958: Second Report to the Universe – Nation of Celestial Space.*

Mani, V.S., 1996. The common heritage of mankind: implications for the legal status of property rights on the Moon and celestial bodies. *Proceedings of the 39th Colloquium on the Law of Outer Space*, p. 31.

Mankiewicz, R., 1968. Interventions with respect to permanent stations on the Moon. *Proceedings of the 11th Colloquium on the Law of Outer Space*, p. 163.

Manzini, E., 2002. *Scenarios of Sustainable Ways of Living – Local and Global Visions* [online]. INDACO: Politecnico di Milano. Available at: http://www.dis.polimi.it/manzini-papers/04.01.01-Scenarios-of-sustainable-ways-of-living.doc [accessed 18 September 2006].

Marburger, J., 2006. Keynote address. *44th Robert H. Goddard Memorial Symposium*, 15 March, Greenbelt, Maryland.

Marko, D.E., 1992–93. A kinder, gentler Moon Treaty: a critical review of the current Moon Treaty and a proposed alternative. *Journal of Natural Resources and Environmental Law*, Vol. 8, p. 293.

Markoff, M.G., 1972. Accords particuliers et droit international general de l'espace. *Proceedings of the 15th Colloquium on the Law of Outer Space*, p. 67.

Markoff, M.G., 1974. Implementing the contractual obligation of Art. I, Par. 1 of the Outer Space Treaty 1967. *Proceedings of the 17th Colloquium on the Law of Outer Space*, p. 36.

Marshall, A., 1995. Development and imperialism in space. *Space Policy*, Vol. 11, p. 41.

Martinez, V.H., 1998. La propiedad de los meteoritos. *Diario Norte*, 28 June.

Marx, K., Engels, F., 1846. *Circular Against Kriege.*

Marx, K., Engels, F., 1848. *Manifesto of the Communist Party*. London.

Marx, K., 1875. *The Criticism of the Gotha Program.*

Matte, N.M. (Ed.), 1984. *Space Activities and Emerging International Law.* McGill University, Montreal.

McDougall, W.A., 1985. *The Heavens and the Earth – A Political History of the Space Age*. New York: Basic Books.

Menthe, D., 1998. Jurisdiction in cyberspace: a theory of international spaces. *Michigan Telecommunications and Technology Law Review*, Vol. 4, p. 69.

Merges, R.P., Reynolds, G.H., 1997. Space resources, common property, and the collective action problem. *New York University Environmental Law Journal*, Vol. 6, p. 107.

Meyer, N., 2000. *Introduction to Property Rights: A Historical Perspective*. LGIEN Fact Sheet 2000–2006, University of Illinois Extension.

Micklethwait, B., 1981. Anarchy versus anarchocapitalism. *Libertarian Alliance Political Notes* No. 4.

Miklody, M., 1979. Some remarks on the question of the rights of possession of mineral resources of the celestial bodies. *Proceedings of the 22nd Colloquium on the Law of Outer Space*, p. 179.

Mom, R., 2005. *Law of the Sea Maritime Boundaries and Dispute Settlement Mechanisms*. Working Paper, The United Nations – The Nippon Foundation of Japan Fellowship Programme, December 2005.

Moon Acres, 1974. Advert. *Parade Magazine*. 24 November, p. 16.

Moore, A.L., 1978. Space stations and habitats. *Proceedings of the American Society of International Law*, Vol. 72, p. 268.

Morris, V., Bourloyannis-Vrailas, M.C., 1996. The work of the Sixth Committee at the Fiftieth Session of the UN General Assembly. *American Journal of International Law*, Vol. 90, p. 500.

Myers, D.S., 1975. Political considerations on some aspects of the law of outer space. *Proceedings of the 18th Colloquium on the Law of Outer Space*, p. 66.

Myers, D.S., 1979. Relations with alien intelligences: legal aspects and political realities. *Proceedings of the 22nd Colloquium on the Law of Outer Space*, p. 181.

Myers, D.S., 1980. The Moon Treaty in Legal and Political Perspective. *Proceedings of the 23rd Colloquium on the Law of Outer Space*, p. 49.

NASA, 2005. *Using NASA imagery and linking to NASA web sites* [online], 13 October. Available from: http://www.nasa.gov/audience/formedia/features/MP_Photo_Guidelines.html [accessed 17 September 2006].

National Space Society, 1999. *NSS Statement of Philosophy* [online], 4 December. Available from: http://www.nss.org/about/philosophy.html [accessed 17 September 2006].

Nauges, H., 1979. Legal aspects. *Proceedings of the 22nd Colloquium on the Law of Outer Space*, p. 269.

Naysmith, S., 2001. Star wars in world of real estate. *Sunday Herald*, 28 January.

Nixon, R., 1973. *Letter from Richard Nixon to Foreign Heads of State*, [online]. March 21. Available from: http://www.geocities.com/Area51/Crater/2029/Nixonltr.htm [accessed 23 May 2000].

North, O., 2005. The Trojan horse on America's shores. *Freedom Alliance*, [online]. March 31. Available from: http://www.freedomalliance.org/view_article.php?a_id=546 [accessed 17 September 2006].

North, P.M., 1979. *Cheshire and North Private International Law.* 10th Ed. London: Butterworths.

Ocilla Star, 1937. Records claims to heavens, Moon and stars. *Ocilla Star*, 15 July.

O'Donnell, D.J., 1996. Benefit sharing: the municipal model. *Proceedings of the 39th Colloquium on the Law of Outer Space*, p. 151.

O'Donnell, D.J., Goldman, N., 1997. Astro law as lex communis spatialis. *Proceedings of the 40th Colloquium on the Law of Outer Space*, p. 322.

O'Donnell, D.J., 1998a. A new institution is proposed to manage space resources: the Metanation in space. *Proceedings of the 41st Colloquium on the Law of Outer Space*, p. 23.

O'Donnell, D.J., 1998b. An archenemy revisited: the 1979 Moon Treaty. *'Space 98' Conference*, April 1998, Albuquerque, New Mexico.

O'Donnell, D.J., Harris, P.R., 1998. A lunar economic development authority. *Space Forum*, Vol. 2, p. 271.

O'Donnell, D.J., 1999. Comity in space. *Proceedings of the 42nd Colloquium on the Law of Outer Space*, p. 179.

Okolie, C.C., 1980. Legal interpretation of the 1979 United Nations Treaty concerning the activities of sovereign States on the Moon and other celestial bodies within the meaning of Common Heritage of Mankind. *Proceedings of the 23rd Colloquium on the Law of Outer Space*, p. 61.

Oosterlinck, R., 1996. Tangible and intangible property in outer space. *Proceedings of the 39th Colloquium on the Law of Outer Space*, p. 271.

O'Rourke, P.J., 1996. *American Spectator's Enemies List*. New York: Atlantic Monthly Press.

Ottewell, G., 1989. *The Thousand-Yard Model or the Earth as a Peppercorn* [online]. Greenville: Furman University. Available from: http://www.noao.edu/education/peppercorn/pcmain.html [accessed 17 September 2006].

Pace, S., 1998. Policy options for recognising non-terrestrial property rights. *NSS Conference 'Property Rights and Commercial Space Development'*, 2 December, Washington.

Paxson, E.W., 1993. Sharing the benefits of outer space exploration: space law and economic development. *Michigan Journal of International Law*, Vol. 14, p. 487.

Picarel, M.E., 1969. Algunas consideraciones sobre el producto lunar. *Proceedings of the 12th Colloquium on the Law of Outer Space*, p. 189.

Pinto, C., 2004. El lado oscuro de la luna – El cuento del tío capitulo 8. *TVN* [online], 29 April. Available from: http://www.tvn.cl/programas/elcuentodeltio/2004/capitulo8.asp [accessed 17 September 2006].

Platoff, A.M., 1992. *Where No Flag has Gone Before: Political and Technical Aspects of Placing a Flag on the Moon*. 11 November (NASA Contractor Report 188251).

Plutarchus, c. 46-120. *De Tranquillitate Animi.*

Power, J.P., 1995. Reinvigorating natural resource damage actions through the public trust doctrine. *New York University Environmental Law Journal*, Vol. 4, p. 418.

Pryor, A., 2003. *Little Known Tales in Nevada History*. Roseville: Stagecoach Publishing.

Quadri, Rolando, 1959. Droit International Cosmique. *98-III Recueil des Cours de l'Academie de Droit International*, p. 505.

Rao, K.N., 1981. Common heritage of mankind and the Moon Treaty. *Indian Journal of International Law*, Vol. 21, p. 275.

Rayment, W.J., 2000. *Analysis of the Communist Manifesto*, [online]. Available from: http://www.indepthinfo.com/communist-manifesto/analysis.shtml [accessed 17 September 2006].

Reid, K.G.C., 1996. *The Law of Property in Scotland.* Edinburgh: The Law Society of Scotland/Butterworths.

Reisenberg, M., 1952. The Moon is yours. *10th Anniversary Science Fiction Convention Bulletin* 1, p. 8.

Reno Evening Gazette, 1929. City bunco men's games fail in country. *Reno Evening Gazette*, p. 9.

Reuter, 1956. They asked for the Moon. *The Gleaner*, 8 April, p. 13.

Reuters, 1997. Get off Mars, it's ours, Yemeni men tell US. *ThinkQuest*, [online]. 24 July. Available from: http://library.thinkquest.org/11967/yemeni.html [accessed 17 September 2006].

Reynolds, G.H., 1995. The Moon Treaty: prospects for the future. *Space Policy*, Vol. 11, p. 115.

Rhodes, B., 1955. For sale: plots out of this world. *Newsday*, 16 November.

Rhodes, C.J., 1902. *Last Will and Testament.*

RIA Novosti, 2005. Russian astrologist loses damages claim against NASA. *RIA Novosti*, [online]. 8 November. Available from: http://en.rian.ru/russia/20051108/42022896.html [accessed 17 September 2006].

Richelson, J.T., 2000. Shooting for the Moon. *Bulletin of the Atomic Scientists*, [online]. September/October. Available from: http://www.space4peace.org/articles/richelson.htm [accessed 19 May 2008].

Robbins, R.M., 1942. *Our Landed Heritage: The Public Domain, 1776–1936.* Princeton: Princeton University Press.

Roberts, L.D., 1990. The law of the commons: framework for the efficient and equitable use of the Lagrange Points. *Connecticut Journal of International Law*, Vol. 6, p. 151.

Roberts, L.D., Pace, S., Reynolds, G.H., 1996. Playing the commercial space game: time for a new rule book? *Ad Astra,* May/June.

Robson, P., Miller, K., 1991. *Property Law.* Edinburgh: W. Green/Sweet & Maxwell.

Roby, H.J., 1902. *Roman Private Law in the Times of Cicero and of the Antonines*. Vol. I. Cambridge: Cambridge University Press.

Rodriguez, G., 2004. Telepossession transforms asteroids into resources. *Space Resources Roundtable VI, Colorado School of Mines*, [online]. Available from http://www.higp.hawaii.edu/srr/SRR-VI-presentations/Rodriguez-TelePoss_SRR6.pdf [accessed 3 May 2008].

Rojas, M.C., 2004. Un chileno dio el permiso. *El Mercurio*, 20 July.

Rosenfield, S.B., 1977. Private industrial development in outer space. *Proceedings of the 20th Colloquium on the Law of Outer Space*, p. 440.

Rosenfield, S.B., 1980. A Moon Treaty? Yes, but why now? *Proceedings of the 23rd Colloquium on the Law of Outer Space*, p. 69.

Rothblatt, M., 1983. A legal charter for non-governmental space industrialization. *Advances in the Astronautical Sciences*, Vol. 53, p. 229.

Rubin, A.P., 1997. *Ethics and Authority in International Law*. Cambridge: Cambridge University Press.

Rusconi, F.G., 1969. Regime of the property of the natural resources on the Moon and other celestial bodies. *Proceedings of the 12th Colloquium on the Law of Outer Space*, p. 185.

Saint-Germain, M., 1997. Pour quelques arpents de lune.... *Radio Canada*, [online]. 14 July. Available from: http://radio-canada.ca/radio/clairetnet/140997/1_2.htm [accessed 13 January 2000].

Schmidt, M.G., 1989. *Common Heritage or Common Burden? The United States Position on the Development of a Regime for Deep Sea-bed Mining in the Law of the Sea Convention.* Oxford: Clarendon Press.

Schmitt, D.G., 2001. Law of ownership and control of meteorites. *The 64th Annual Meteoritical Society Meeting*.

Scripps Howard News Service, 1998. Own your piece of the Rock. *San Francisco Examiner*, 21 February.

Seward, W.H., 1850. *The "Higher Law" Speech*, [online]. March 11. Available from: http://facweb.furman.edu/~benson/docs/seward.htm [accessed 17 September 2006].

Shiffer, H.A., Snyders, P.W., 1967. The need for enforcement for international space treaties. *Proceedings of the 10th Colloquium on the Law of Outer Space*, p. 237.

Silber, K., 1998. A little piece of heaven – space-based commercial development will happen sooner than you think. How a system of extraterrestrial property rights might emerge. *Reason*, November. Available from: http://www.reason.com/news/show/30796.html

Silving, H., 1968. *Sources of Law.* Buffalo: Hein.

Simberg, R., 2004. The path not taken. *The New Atlantis,* Number 6, p. 46.

Simpson, S.R., 1976. *Land Law and Registration.* Cambridge: Cambridge University Press.

Smirnoff, M., 1966. Introductory report and summary of discussions – draft resolution on the legal status of celestial bodies. *Proceedings of the 9th Colloquium on the Law of Outer Space*, p. 8.

Smirnoff, M., 1972. The need for a treaty on the legal status of the Moon. *Proceedings of the 15th Colloquium on the Law of Outer Space*, p. 73.

Smith, B., Zaibert, L., 1997. *The Metaphysics of Real Estate*, [online]. University of Buffalo. Available from: http://wings.buffalo.edu/academic/department/philosophy/faculty/smith/articles/lz.htm [accessed 3 December 1999].

Smith, M.L., 1988. The commercial exploitation of mineral resources in outer space. In: T.M. Zwaan & al, (Eds.) *Space Law: Views of the Future*. Deventer: Kluwer Law and Taxation Publishers, p. 45.

Sonter, M.J., 1998. The technical and economic feasibility of mining the near-earth asteroids. *The 49th IAF Congress*, September 28–October 2, Melbourne.

Sowell, T., 1999. *Barbarians Inside the Gates – and Other Controversial Essays.* Stanford: Hoover Institution Press.

Stair, J., 1693. *The Institutions of the Law of Scotland*. Edingurgh.

Stardate, 1995. *How Fast is Earth Moving Through Space*?, [online]. Available from: http://stardate.org/resources/faqs/008.html [accessed 17 September 2006].

Statesville Landmark, 1946. Makes application for land on Moon. *The Statesville Landmark*, 17 October, p. 1.

Steinberg, T., 1996. *Slide Mountain – Or the Folly of Owning Nature*. Reprint edition. Berkeley: University of California Press.

Sterns, P.M., Tennen, L.I., 1979. International recognition of "the art of living in space": the emergence of settlement competence. *Proceedings of the 22nd Colloquium on the Law of Outer Space*, p. 221.

Sterns, P.M., Tennen, L.I., 1992. Utilization of extraterrestrial resources: law, science and policy. *Proceedings of the 35th Colloquium on the Law of Outer Space*, p. 499.

Sterns, P.M., Tennen, L.I., 1999. Should the lunar crater SAHA be accorded special legal protection? *Proceedings of the 42nd Colloquium on the Law of Outer Space*, p. 393.

Sterns, P.M., Tennen, L.I., 2002. Privateering and profiteering on the Moon and other celestial bodies: debunking the myth of property rights in space. *World Space Congress*, Houston (COSPAR 02-A-01933).

Sterns, P.M., Stine, G.H., Tennen, L.I., 1996. Preliminary jurisprudential observations concerning property rights on the moon and other celestial bodies in the commercial space age. *Proceedings of the 39th Colloquium on the Law of Outer Space*, p. 50.

Stevens Point Daily Journal, 1965. Want to claim some lunar land. *Stevens Point Daily Journal*, 7 May, p. 14.

Stirner, M., 1844. *The Ego and its Own*. Leipzig.

Strong, M.F., 1995. *Post-Rio Sustainable Development and the Summit*, [online]. May 22. Available from: http://www.library.utoronto.ca/www/g7/rt95sgl.htm [accessed 10 February 1999].

Stumpf, B.G., 1998. *Africa on the Matrix: The Masai*, [online]. Available from: http://www.on-the-matrix.com/africa/masai_people.asp [accessed 17 September 2006].

Suburbanite Economist, 1970. 'Celestia' loses Prime Minister. *Suburbanite Economist*, 19 July.

Sweet, C., 1882. *A dictionary of English Law*. London: Sweet.

Szaloky, L., 1969. Activities on celestial bodies including the exploitation of possible natural resources there. *Proceedings of the 12th Colloquium on the Law of Outer Space*, p. 76.

Szaloky, L., 1973. The way of the further perfection of the legal regulation concerning the Moon and other celestial bodies, especially regarding the exploitation of natural resources of the Moon and other celestial bodies. *Proceedings of the 16th Colloquium on the Law of Outer Space*, p. 196.

Sztucki, J., 1966. Remarks during the discussion on the introductory report. *Proceedings of the 9th Colloquium on the Law of Outer Space*, p. 64.

Tamm, J.R., 1968. Legal status of national stations on celestial bodies. *Proceedings of the 11th Colloquium on the Law of Outer Space*, p. 158.

Tank, R.W., 1983. *Legal Aspects of Geology*. New York and London: Plenum Press.

TASFIC, 1952. Moon titles secure despite rival claims. *10th Anniversary Science Fiction Convention Bulletin*, no.2, p. 8.

The Archimedes Institute, 1997a. *Real Property Registry*, [online]. Available from: http://www.permanent.com/archimedes/PropertyRegistry.html [accessed 17 September 2006].

The Archimedes Institute, 1997b. *Claim Registration*, [online]. Available from: http://www. permanent.com/archimedes/register.htm [accessed 17 September 2006].

The Chronicle Telegram, 1953. Whose Moon? *The Chronicle Telegram*, 2 June, p. 28.

The Daily Northwestern, 1910. Wants square deal. *The Daily Northwestern*, 6 December, p. 5.

The Sheboygan Press, 1971. Do you remember? *The Sheboygan Press*, 27 August.

The Texas Docket, 2000. Space law: national and individual rights in outer space. *The Texas Docket*, April.

Thomas, C.D. (Reporter), 1980. The Moon Treaty: Should the United States become a party? *Proceedings of the American Society of International Law*, Vol. 74, p. 152.

Thomas, J.A.C., 1975. *The Institutes of Justinian – Text, Translation and Commentary*. Amsterdam and Oxford: North-Holland Publishing Company.

TransOrbital, 2002. *TrailBlazer Text Message*, [online]. Available from: https://www.transorbital.net/catalog/text_message.php [accessed 17 September 2006].

Tuchman, B., 1976. On our birthday—America as idea. *Newsweek*, 12 July.

Tumlinson, R., 2003. *Testimony before the Senate Committee on Commerce, Science and Transportation*, October 29, p. 9.

Turner, F.J., 1893. The significance of the frontier in American history. *Proceedings of the State Historical Society of Wisconsin*, 14 December.

United Nations General Assembly, 1995. *Trusteeship Council Proposals Range from Disbandment to New Mandate, as General Assembly Debates Council's Future.* 21 November (Press Release GA/9008).

United Press, 1949. 'Owner' of Space fears trespasser. *Nebraska State Journal*, 14 August, p. 6A.

United Press, 1957. He'd give up deed to Space. *The Hammond Times*, 15 October, p. A4.

United Press International, 1965. Lots on Moon for sale.*The Oakland Tribune*, 18 July.

United Press International, 1967. Soviets poaching on Texas town's suburb. *The Stars and Stripes*, 24 October, p. 6.

United Press International, 1969. Moon sale was going good. *The Valley Independent*, 16 August, p. 9.

United States Senate, 1980. *Hearings before the Subcommittee on Science, Technology, and Space of the Committee on Commerce, Science, and Transportation, United States Senate, Ninety-Sixth Congress, Second Session, on Agreement Governing the Activities of States on the Moon and Other Celestial Bodies*. July 29 and 31, 1980. Washington: US Government Printing Office (Serial No. 96-115).

Urbahn, K., 2005. Putting the torch to a culture of entitlement. *Yale Daily News*, 23 March. Available from: http://www.yaledailynews.com/articles/view/13693

US Bureau of Land Management, 2000. *Does the Federal Government Ever Sell Public Land?* [online]. Available from: http://www.blm.gov/nhp/what/lands/realty/pubs/land4sal/index.htm [accessed 17 September 2006].

US Department of State, 1974. Statement on claim of exclusive mining rights by Deepsea Ventures, Inc. *International Legal Materials*, Vol. 14, p. 66.

Van Traa-Engelman, H.L., 1980. The Moon Treaty – legal consequences and practical aspects. *Proceedings of the 23rd Colloquium on the Law of Outer Space*, p. 73.

Vassilevskaya, E.G., 1977. Notions of 'exploration' and 'use' of natural resources of celestial bodies. *Proceedings of the 20th Colloquium on the Law of Outer Space*, p. 473.

Vassilevskaya, E.G., 1980. What is new in the legal regulations of activities in outer space? *Proceedings of the 23rd Colloquium on the Law of Outer Space*, p. 79.

Vassilevskaya, E.G., 1984. On the setting up of large structures on the Moon. *Proceedings of the 27th Colloquium on the Law of Outer Space*, p. 273.

Vicas, A.G., 1980. The New International Economic Order and the emerging space regime. In: *Space Activities and Implications: Where from and Where to at the Threshold of the 80's, Proceedings of the Symposium held on October 16–17, 1980*. Montreal: McGill University, p. 293.

Vinogradoff, P., 1959. *Common Sense in Law*. New York: Oxford University Press.

Von der Dunk, F., 2000. *Question and Answer*, [online]. Available from: http://ruljis.leidenuniv.nl/group/jflr/www/Bulltn.Board/Q-and-A._Ownership_of_the_ Moon.htm> [accessed 11 May 2001].

Von der Dunk, F.G., 1996. The role of law with respect to future space activities. *Space Policy*, Vol. 12, p. 5.

Von der Dunk, F.G., 1999. Earth observation and data policy in Europe: the legal issues – the EOPOLE concerted action project. *Proceedings of the 42nd Colloquium on the Law of Outer Space*, p. 373.

Von der Dunk, F.G., 2007. Back in Business? The Moon Agreement, Private Actors and Possible Commercial Exploitation of the Moon and Its Natural Resources, in *Proceedings of the International and Interdisciplinary Workshop on Policy and Law Relating to Outer Space Resources: Examples of the Moon, Mars, and Other Celestial Bodies*. Montreal: McGill University Center for Research of Air and Space Law, p. 244.

Von Rauchhaupt, F.W., 1969. Las fuentes principales del derecho especial. *Proceedings of the 12th Colloquium on the Law of Outer Space*, p. 201.

Wadegaonkar, D., 1984. *The Orbit of Space Law*. London: Stevens & Sons/Bombay: N.M. Tripathi.

Wassenbergh, H.A., 1991. *Principles of Outer Space Law in Hindsight*. Dordrecht: Martinus Nijhoff Publishers.

Wasser, A., 2004a. *(Draft of) An Act to Promote Privately Financed Space Settlement*, [online]. Available from: http://www.spacesettlement.org/law/ [accessed 17 September 2006].

Wasser, A., 2004b. *The Space Settlement Initiative*, [online]. Available from: http://www. spacesettlement.org [accessed 17 September 2006].

Webber, A.D., 1983. Extraterrestrial law on the final frontier: a regime to govern the development of celestial body resources. *Georgetown Law Journal*, Vol. 71, p. 1427.

Westfall, R.M., 2003. Establishing possession of debris and resources in space. *Galactic Mining Industries*, [online]. Available from http://www.angelfire.com/trek/galactic_mining/rmwgmi.html [accessed 6 May 2008].
Westfall, R.M., O'Donnell, Declan J., Rodriguez, G., 2003. Telepossession of extraterrestrial resources and leveraged financing of outer space projects. *Space Resources Roundtable V, Colorado School of Mines*, [online]. Available from http://www.isruinfo.com/docs/srr5/westfall_et_al.zip [accessed 3 May 2008].
White, W.N., 1985. *Real Property Rights in Outer Space.* Unpublished academic version.
White, W.N., 1991. Mining Law for outer space. *Space Manufacturing*, Vol. 8, p. 83.
Whitehouse, D., 2001. Planet Earth on the move. *BBC News*, [online], 5 February. Available from: http://news.bbc.co.uk/hi/english/sci/tech/newsid_1154000/1154784.stm [accessed 17 September 2006].
Whiteman, M.M. (Ed.), 1964. *Digest of International Law.* Vol. 3. Washington, DC: Department of State.
Whiteman, M.M. (Ed.), 1965. *Digest of International Law*, Vol. 4. Washington, DC: Department of State.
Wiewiorowska, K., 1979. Legal and political problems of the solar power stations in space. *Proceedings of the 22nd Colloquium on the Law of Outer Space*, p. 23.
Wiewiorowska, K., 1980. Implications of the Moon Agreement for the legal status of outer space. *Proceedings of the 23rd Colloquium on the Law of Outer Space*, p. 83.
Wikipedia contributors 2006. Wilderness hut. *Wikipedia*, [online]. Available from: http://en.wikipedia.org/w/index.php?title=Wilderness_hut&oldid=57640844 [accessed 6 June 2006].
Wiles, G.E., 1998. The Man on the Moon makes room for neighbors: an analysis of the existence of property rights on the Moon under a condominium – type ownership theory. *International Review of Law Computers and Technology*, Vol. 12, p. 513.
Williams, S.M., 1969. Utilization of meteorites and celestial products. *Proceedings of the 12th Colloquium on the Law of Outer Space*, p. 179.
Williams, S.M., 1987. The law of outer space and natural resources. *International and Comparative Law Quarterly*, Vol. 36, p. 142.
Williams, S.M., 1997. The development of Article IX of the 1967 Space Treaty. *Proceedings of the 40th Colloquium on the Law of Outer Space*, p. 177.
Woetzel, R.K., 1962. Sovereignty and national rights in outer space and on celestial bodies. *Proceedings of the 5th Colloquium on the Law of Outer Space.*
Zhukov, G.P., 1967. The problem of the definition of outer space. *Proceedings of the 10th Colloquium on the Law of Outer Space*, p. 271.
Zubrin, R., 1999. *Entering Space: Creating a Spacefaring Civilization.* Tarcher/Putnam, New York.
Zunt, D., 1998. Who did actually invent the word "robot" and what does it mean? *The Karel Čapek Website*, [online]. Available at http://capek.misto.cz/english/robot.html [accessed 2 May 2008].
Zwicky, F., 1972. Examples of activities in extraterrestrial space which might be judged harmful, harmless, useful, or either one of these, depending on the viewpoint. *Proceedings of the 15th Colloquium on the Law of Outer Space*, p. 259.

Table of Cases

City of Milwaukee v. State [1927]. Supreme Court of Wisconsin. 193 Wis. 423.
Columbus-America Discovery Group, Inc. v. The Unidentified, Wrecked and Abandoned Sailing Vessel, S.S. Central America [1989] A.M.C. 1955 (E.D. Va. 1989).
Geer v. Connecticut [1896]. Supreme Court of the United States. 161 U.S. 519.
Goddard v. Winchell [1892]. Supreme Court of Iowa. 86 Iowa 71, 52 N.W. 1124.
Island of Palmas Case [1928]. Scott, Hague Court Reports 2d 83 (1932) (Perm. Ct. Arb. 1928), 2 U.N. Rep. Intl. Arb. Awards 829.

Jacobsen v. Norwegian Government [1933]. Supreme Court of Norway. Annual Digest and Reports of Public International Law Cases; 1933–34: case no. 42.
Johnson v. M'Intosh [1823]. Supreme Court of the United States. 21 U.S. 543.
Minquiers and Ecrehos Case [1953]. International Court of Justice.
Nemitz v. United States [2004]. U.S. District Court, Nevada.
Oregon Iron Co. v. Hughes [1905]. Supreme Court of Oregon. Pacific Reporter, 579.
The Case of the S.S. "Lotus" [1927]. Permanent Court of International Justice. Judgement No. 9.
The Kronprinsessan Margareta [1921]. Privy Council. 1 A.C. 486 PC.

Index

FSC
www.fsc.org
MIX
Papier | Fördert gute Waldnutzung
FSC® C083411

Kolibri 360 GmbH
Ferdinand-Jühlke-Straße 7
99095 Erfurt, Deutschland
produktsicherheit@kolibri360.de